Translational Bioinformatics

Volume 5

Aims and Scope

The Book Series in Translational Bioinformatics is a powerful and integrative resource for understanding and translating discoveries and advances of genomic, transcriptomic, proteomic and bioinformatic technologies into the study of human diseases. The Series represents leading global opinions on the translation of bioinformatics sciences into both the clinical setting and descriptions to medical informatics. It presents the critical evidence to further understand the molecular mechanisms underlying organ or cell dysfunctions in human diseases, the results of genomic, transcriptomic, proteomic and bioinformatic studies from human tissues dedicated to the discovery and validation of diagnostic and prognostic disease biomarkers, essential information on the identification and validation of novel drug targets and the application of tissue genomics, transcriptomics, proteomics and bioinformatics in drug efficacy and toxicity in clinical research.

The Book Series in Translational Bioinformatics focuses on outstanding articles/chapters presenting significant recent works in genomic, transcriptomic, proteomic and bioinformatic profiles related to human organ or cell dysfunctions and clinical findings. The Series includes bioinformatics-driven molecular and cellular disease mechanisms, the understanding of human diseases and the improvement of patient prognoses. Additionally, it provides practical and useful study insights into and protocols of design and methodology.

Series Description

Translational bioinformatics is defined as the development of storage-related, analytic, and interpretive methods to optimize the transformation of increasingly voluminous biomedical data, and genomic data in particular, into proactive, predictive, preventive, and participatory health. Translational bioinformatics includes research on the development of novel techniques for the integration of biological and clinical data and the evolution of clinical informatics methodology to encompass biological observations. The end product of translational bioinformatics is the newly found knowledge from these integrative efforts that can be disseminated to a variety of stakeholders including biomedical scientists, clinicians, and patients. Issues related to database management, administration, or policy will be coordinated through the clinical research informatics domain. Analytic, storage-related, and interpretive methods should be used to improve predictions, early diagnostics, severity monitoring, therapeutic effects, and the prognosis of human diseases.

Recently Published and Forthcoming Volumes

Applied Computational Genomics
Editor: Yin Yao Shugart
Volume 1

Pediatric Biomedical Informatics
Editor: John Hutton
Volume 2

Bioinformatics of Human Proteomics
Editor: Xiangdong Wang
Volume 3

Bioinformatics for Diagnosis, Prognosis and Treatment of Complex Diseases
Editor: Bairong Shen
Volume 4

More information about this series at http://www.springer.com/series/11057

Xiangdong Wang
Editor

Single Cell Sequencing and Systems Immunology

Honor editors
Xiaoming Chen
Zhihong Sun
Jinglin Xia

Springer

Editor
Xiangdong Wang
Zhongshan Hospital
Fudan University
China

Shanghai Institute of Clinical Bioinformatics
China

ISSN 2213-2775 ISSN 2213-2783 (electronic)
Translational Bioinformatics
ISBN 978-94-017-9752-8 ISBN 978-94-017-9753-5 (eBook)
DOI 10.1007/978-94-017-9753-5

Library of Congress Control Number: 2015936190

Springer Dordrecht Heidelberg New York London

Printed on acid-free paper

Springer Science+Business Media B.V. Dordrecht is part of Springer Science+Business Media (www.springer.com)

Preface

The Volume of ***Single Cell Sequencing and Systems Immunology*** has taken more than a year to prepare and write, due to the rapid development of science and discovery. This particular volume brings a special attention on practical application of single cell sequencing and systems immunology for clinical research or human materials-based studies. It has been a true labour of protocol development, including methodological design and collection, practical repeatability, and handling experience. It has been a hard time of concept definition and clarification of systems immunology. We hope that the readers can find the link between single cell sequencing and systems immunology, apply single cell sequencing for the understanding of systems immunology, and integrate both to discover new disease-specific biomarkers and therapeutic targets. The most important feature of the Volume is to emphasize the potential of single cell biology and function in the pathogenesis of diseases. It is a time to explore alterations of human single cell biological function, epigenetics, metabolisms, or gene/protein networks and interactions. The Volume is particularly suitable for biomedical students, practical junior researchers, and principle investigators who are working and planning to perform studies on single cell sequencing and systems immunology. One of the greatest joys in creating this Volume has been working with all outstanding experts, especially those young researchers and scientists who gave this project not just their time and effort, but their hearts and souls. Their seriousness and hard work deserved the highest respect and congratulation. I do not believe we can create such exciting and stimulating Volume without their dedication and contributions.

China — Xiangdong Wang

Contents

1 **Significance of Single Cell Sequencing in Future Medicine** 1
Mengjia Qian, Duojiao Wu, Ding Zhang, Jian Wang, Yunfeng Cheng, and Xiangdong Wang

2 **Uncovering Phenotypes with Supercells: Applications to Single-Cell Sequencing** 11
Julián Candia, Jayanth R. Banavar, and Wolfgang Losert

3 **Single-Cell Sequencing of Microorganisms** 31
Ying Chen and Fengping Wang

4 **The Application of Single-Cell Sequencing in Dynamic Transcriptomes** 41
Kenian Chen, Xiaomin Dong, and Jiaqian Wu

5 **System Immunology in Metabolic Research of Dendritic Cells** 65
Duojiao Wu

6 **$FOXP3^+$ Treg Cells and Systems Biology Approaches to Studying Their Function** 73
Xuerui Luo, Andy Tsun, Jing Yang, Zuojia Chen, Yayi Gao, Kongchen Li, Miranda Piccioni, and Bin Li

7 **$\gamma\delta$T Cells and $\alpha\beta$T Cells** 95
Lingyan Wang, Fuping Wang, Jieyi Shi, and Xiangdong Wang

8 **Prolymphocyte: Lymphocytes and Bioinformatics** 111
Mengjia Qian, Claudio Spada, and Xiangdong Wang

9 **Lymphocytes in Liver Cancer** 123
Xiaojing Xu, William C.S. Cho, and Jinglin Xia

10 **Early Biomarkers of Hepatocyte Necrosis** 143
Mitchell R. McGill and Hartmut Jaeschke

11 Protocol for Single Cell Isolation by Flow Cytometry 155
Jian Wang, Zhihui Min, Meiling Jin, and Xiangdong Wang

12 A Simple Protocol for Single Lung Cancer Cell Isolation-Making the Single Cell Based Lung Cancer Research Feasible for Individual Investigator 165
Ding Zhang and Xiangdong Wang

13 Cancer Bioinformatics 175
Mingyan He, Li Feng, and Jinglin Xia

About the Editors

Xiangdong Wang, M.D., Ph.D., is a Distinguished Professor of Respiratory Medicine at Fudan University, Director of Fudan University Center of Clinical Bioinformatics, Deputy Director of Shanghai Respiratory Research Institute, Adjunct Professor of Clinical Bioinformatics at Lund University, and Visiting Professor of King's College of London. He serves as a Director of Biomedical Research Center, Fudan University, Zhongshan Hospital. His main research is focused on clinical bioinformatics, disease-specific biomarkers, cancer immunology, and molecular and cellular therapies.

His group integrates clinical informatics with omics science and bioinformatics to identify and validate disease-specific biomarkers and therapeutic targets in chronic lung diseases and lung cancer. His group initially developed the mirror-butterfly chemical structure of phosphoinositide 3-kinase inhibitor to prevent and treat chronic lung inflammation and injury, in combination of his pharmaceutical experience of drug discovery and development.

In addition, Dr Wang serves as the Executive Vice President of International Society for Translational Medicine, Chairman of Executive Committee of International Society for Translational Medicine, Deputy President of Chinese National Professional Society of Insurance & Health and a Senior Advisor of Chinese Medical Doctor Association, and Director of National Program of Doctor-Pharmaceutist Communication. Dr Wang was appointed as the Principal Scientist, Global Disease Advisor, Medical Monitor and Director, and Chairman of Director Board in a number of pharmaceutical companies, e.g., Astra Draco, AstraZeneca, PPT, and CatheWill. He worked on pharmacology profiles of target identification and validation, drug screening and optimization, drug PK and PD profile, and translation between discovery and development in areas of respiratory diseases, inflammation, and cancer. He acted as the Adjunct Professor of Molecular Bioscience at

North Carolina State University and the Member of American Thoracic Society International Health Committee, USA.

He serves as Co-editor in chief of *Journal of Clinical Bioinformatics*, *Clinical and Translational Medicine*, and *Molecular and Cellular Therapies*; Editor of *Serial Book, Translational Bioinformatics*; Section Editor of Disease Biomarkers of *Journal of Translational Medicine*; Asian Editor of *Journal of Cellular Molecular Medicine*; and the Editorial Member of international journals, e.g., *American Journal of Critical Care Medicine* and *American Journal of Cellular and Molecular Biology*. He is the author of more than 200 scientific publications with the impact factor about 500, citation number about 2000, and cited journal impact factor about 5000.

Xiaoming Chen, M.D., Professor and Chief Physician of Pediatric Surgery, Deputy President of Wenzhou Medical College, and President of the First Hospital of Wenzhou Medical University. He graduated in Faculty of Clinical Medicine, Wenzhou Medical University, in 1980. He serves as Deputy Supervisor of Oncology in Chinese Medical Association, Director of Chinese Hospital Association, Active Member of Teaching Guidance Committee of Forensic Medical of Ministry of Education, Director of Pediatric Surgery Division of Zhejiang Medical Association, Vice Chairman of Anticancer Association of Zhejiang province, Director of Tumor Biological Therapy Committee of Zhejiang province, and Executive Director of Zhejiang Eugenics Association. He serves as the Executive Director of Wenzhou Red Cross Association, Chairman of Zhejiang Federation of Returned Overseas Chinese Member, Director of Wenzhou People's Association for Friendship with Foreign Countries, and Deputy Director of Wenzhou Education Association for International Exchange. Dr Chen's research areas include pediatric tumors, clinical and basic research on male reproductive system, and hospital management. Monographs published include *Guidelines for Diagnosis and Treatment of Solid Tumors in Children* as a Deputy Editor, *Hospital Performance Management* as a Deputy Editor, and *Baby Don't Cry* as a Chief Editor. More than 20 papers have been published in recent years. Dr Chen developed an integrated system of hospital information administration and management to gather all services, e.g., outpatients, inpatients, operations, staff administration, payment, registration, and others.

Hongzhi Sun, M.D., Ph.D., Professor and Director of Gastroenterological Cancer Center, President of Liaoning Medical University First Affiliated Hospital, Chairman of General Surgical and Biological Materials, Professional Committee of Liaoning Province Society of Cell Biology, Deputy Director of Professional Committee of Cancer Biology Research, Vice Chairman of Biotechnology Committee of Liaoning Province Society, and Executive Director of Liaoning Province Gastric Anticancer Association.

Jinglin Xia, M.D., Ph.D., Deputy Dean of Medical College, President of Minhang Hospital, Fudan University, Shanghai, China. He finished postdoctoral research at the Medical Center of Pittsburgh University in 2002–2003. He serves as the editor of a number of international medical journals, Executive Member of the International Association for Translational Medicine, and Vice Chairman of the Shanghai Medical Association. Dr Xia's main research focuses on tumor angiogenesis and tumor inflammatory microenvironment. Dr Xia has published over 60 papers and plenty of books covering tumor angiogenesis, microRNAs, inflammatory factors, and so on.

Chapter 1
Significance of Single Cell Sequencing in Future Medicine

Mengjia Qian, Duojiao Wu, Ding Zhang, Jian Wang, Yunfeng Cheng, and Xiangdong Wang

Abstract Nowadays, the conception of single cell sequencing has been widely referred, it can be used to develop new diagnostics, monitor disease progresses, measure responses to therapies, and predict the prognosis of patients. In this chapter, we discussed the potential applications, clinical significance, workflow compared with microarray, challenges and perspectives of single cell sequencing.

Keywords Single cell • Transcriptome • Sequencing • Medicine • Bioinformatics

1.1 Introduction

The sequencing as a new breakthrough and emerging field was initiated since 1998 by the first 96-capillary sequencers, which was called high-throughput sequencing and referred as "first-generation technology". Over the past decade, new sequencing technologies have been redefined from the "high-throughput" term to

M. Qian, B.S. • Y. Cheng, M.D., Ph.D. (✉)
Biomedical Research Center, Zhongshan Hospital, Fudan University, Shanghai, China

Shanghai Clinical Bioinformatics Research Institute, Shanghai, China
e-mail: yfcheng@fudan.edu.cn

D. Wu, M.D., Ph.D.
Biomedical Research Center, Zhongshan Hospital, Fudan University, Shanghai, China

Shanghai Institute of Clinical Bioinformatics, Shanghai, China

D. Zhang
Shanghai Institute of Clinical Bioinformatics, Shanghai, China

J. Wang
Shanghai Clinical Bioinformatics Research Institute, Shanghai, China

X. Wang, M.D., Ph.D. (✉)
Zhongshan Hospital, Fudan University, China

Shanghai Institute of Clinical Bioinformatics, China
e-mail: xiangdong.wang@clintransmed.org

X. Wang (ed.), *Single Cell Sequencing and Systems Immunology*,
Translational Bioinformatics 5, DOI 10.1007/978-94-017-9753-5_1

"nextgeneration of sequencing" (Martin et al. 2013). The term of 'next generation of sequencing' has been widely adopted in the near 5 years. Single cell transcriptome defined as the entire RNA or polyadenylated products of RNA polymerase II on a cell can describe the gene regulation networks responsible for physiological functions, behaviors, and phenotypes in response to signals and microenvironmental changes. The single cell research and value will be also dependent upon the understanding of genomic heterogeneity, planning and design of study protocol, representative of selected and targeted cells, and sensitivity and repeatability of the methodology. We recently call special attentions and define opportunities and challenges of single-cell sequencing for clinical application (Zhu et al. 2014). The single cell sequencing can be used to develop new diagnostics, monitor disease progresses, measure responses to therapies, and predict the prognosis of patients, although there are still a large number of challenges and difficulties to be faced. It would be more values and specificities of the single cell sequencing to integrate with the function of cells, organs, and systems of the body, the clinical phenotypes of patients, and the description of clinical bioinformatics. The present chapter will highlight the potential application, significance, workflow, comparison, perspective in future, and challenges of single cell sequencing for discovery and development of disease-specific biomarkers.

1.2 Potential Applications

Single cell transcriptome/sequencing has been strongly suggested to be one of the most important ways to identify and validate disease biomarkers and therapeutic targets (Soon et al. 2013). Single cell sequencing was believed to have the special power to investigate small groups of differentiating cells, circulating tumour cells, or tissue stem cells. The single-cell sequencing has been used to identify genetic changes in rare cells, e.g. cancer and tissue stem cells, in clinical samples. The methodologies of single-cell sequencing methods have been improving and developing with the increase of the understanding and attention. We proposed apply the single cell sequencing analysis for the clinical research and application to identify and validate disease-specific biomarkers, network biomarkers, dynamic network biomarkers. The single cell research and value will be also dependent upon the understanding of genomic heterogeneity, planning and design of study protocol, representative of selected and targeted cells, and sensitivity and repeatability of the methodology.

The single cell sequencing can be used develop new diagnostics, monitor disease progresses, measure responses to therapies, and predict the prognosis of patients, although there are still a large number of challenges and difficulties to be faced. It would be more values and specificities of the single cell sequencing to integrate with the function of cells, organs, and systems of the body, the clinical phenotypes of patients, and the description of clinical bioinformatics. Methods of single cell transcriptome analyses have been developed and significantly improved with advantages and disadvantages (Martin et al. 2013; Soon et al. 2013). Single cell transcriptome can demonstrate the heterogeneity of gene expressions, interactions and regulations of gene regulatory networks, subpopulations within a tumour, characteristics of putative cancer stem cells, gene expression profiles of intracellular compartments, mRNA locations, allele specific gene expression, or the information combined with the existing strand-specific cDNA library preparation strategies. With next generation sequencing, the information on both full genome and transcriptome can be obtained from an individual cell. Single-cell genome analysis becomes more important in the understanding of the diversity in microbial ecology, cancer development, prenatal genetic diagnosis, or human genome structures.

1.3 Clinical Significance

Biological research requires the analysis of tissues, cell populations and the whole organisms. Much variation will occur at the single cell level where understanding of each individual cell is critical for the analysis of the entire system (Soon et al. 2013). The breakthrough of single cell sequencing is of great importance and will help to discover cell lineage relationships and to supplant the blurry notion of marker-based cell types. Single cell epigenomics and proteomics will allow the functional states of individual cells to be analyzed. Those integrated technologies will produce more detailed knowledge which enable high-throughput, multi-dimensional analysis of individual cells (Shapiro et al. 2013).

Most sequencing techniques require DNA or RNA from over 10^5 cells (Metzker 2010). The technique of single cell sequencing improves the quantification of rare cells because it needs less starting sample. Scientists have successfully demonstrated that single cell sequencing can effectively be applied in human pre-implantation embryo cells (Xue et al. 2013), individual sperm cells (Wang et al. 2012), singe neuron (Qiu et al. 2012) and peripheral circulation tumor cells (Ramskold et al. 2012). In embryonic stem cell developmental biology, each cell expresses specific transcriptional programs which define its eventual developmental fate (Sylvester and Longaker 2004). The emergence of single cell sequencing enabled this analysis more in-depth and more comprehensively.

In the issue of cancer research, intra-tumoral heterogeneity in carcinogenesis can promote tumor evolution and hinder personalized therapies (Gerlinger et al. 2012). The occurrence of single cell sequencing can help exploring the small number of circulating tumor cells without the interference of non-cancerous cells and different subclones, so that scientists can understand the intra-tumoral heterogeneity, especially with regard to recurrence or resistance to anti-cancer therapies (Yoshida et al. 2013). This may provide a better view for how tumors develop, expand and progress, and thus for exploring individualized treatments for cancer. Via sequencing, a preliminary analysis indicates the appropriateness of other omics technologies such as proteomics, glycomics or metabolomics. It can be a relatively cheap way to determine the likely interesting subsets of samples which are likely to generate results in other omics technologies (McGettigan 2013).

1.4 Practical Workflow

There are some platforms already been tested their efficiency in single cell sequencing in laboratory such as the semiconductor sequencers (Ion Torrent Personal Genome Machine (PGM) and Ion Torrent Proton, Life Technology) which can efficiently perform single cell sequencing at low cost with 3–4 day's turnaround time from cell sampling to sequencing and initial annotation (Kohn et al. 2013). The workflow of single cell sequencing is generally cell isolation, library construction, sequencing and data analysis, as shown in Fig. 1.1.

The single cell can be isolated by various methods we can use such as biochemical purification, fluorescence sorting, laser-assisted microdissection or micromanipulation methods (Wuest et al. 2013). According to different platforms, there are different methods from constructing libraries to performing sequencing. Here, we just introduce a basal workflow via Ion Torrent Sequencing Platform (Fig. 1.2). After construction of sequencing libraries according to different Ion Torrent Library Preparation protocols based on different sample types (Fig. 1.3), each individual DNA fragment will combine with an Ion Sphere Particle (a kind of bead) and clonally amplified automatically on the Ion OneTouch™ System. Then the empty beads (do not combine the DNA fragment) will be eliminated while the amplified, individually cloned DNA fragments will be enriched. This process also performs automatically on a robotic enrichment system (Ion OneTouch™ ES). Finally, the beads containing clonal populations of DNA will be loaded on a sequencing chip, arrayed in wells and incubated serially with pure nucleotides of DNA. Incorporation of a nucleotide is continuously detected by measuring changes in the hydrogen ion concentration during the sequencing process on the sequencer and the data will simultaneously be processed on a server for further analysis (Kohn et al. 2013).

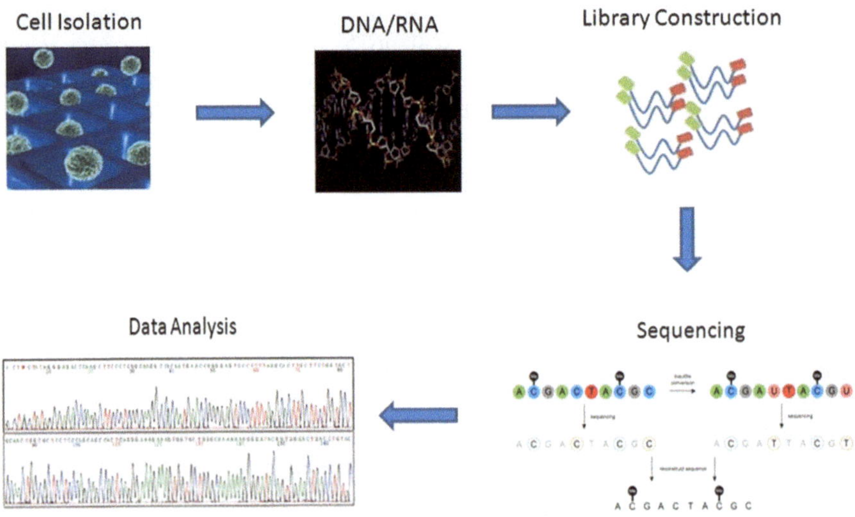

Fig. 1.1 The general workflow of single cell sequencing

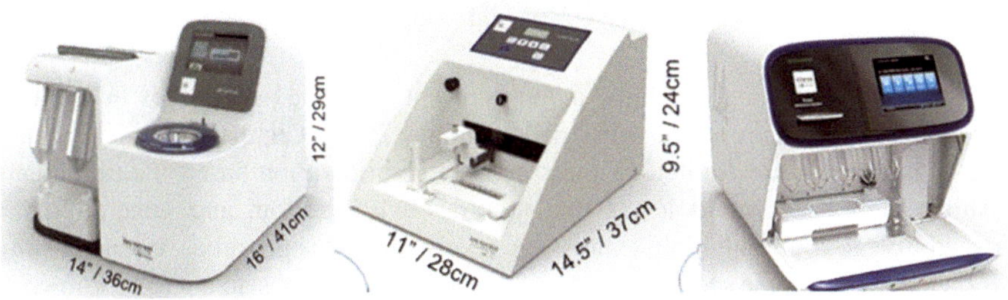

Fig. 1.2 Ion Torrent Sequencing Platform from *left* to *right*, there are Ion OneTouch™ System, Ion OneTouch™ ES, Ion Proton™ System (Sequencer)

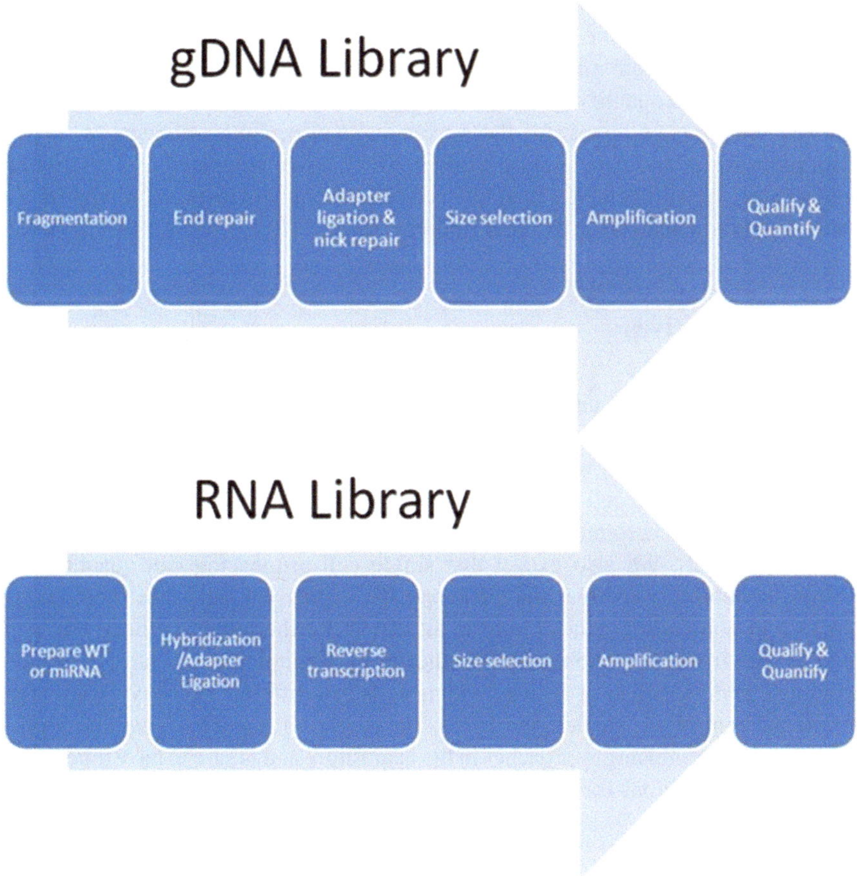

Fig. 1.3 Library preparation workflow

1.5 Comparison with Microarray

Single cell sequencing is unique among current genomic approaches in yielding access to the genomes of individual cells (Blainey 2013). Comparing with sequencing, microarrays are limited in their design. Hybridization-based measurements exhibit high background levels due to cross-hybridization, and generally lack sensitivity at low and high expression levels. The design relies upon existing knowledge in genome sequence, but in sequencing, sequences do not to be known in advance (Wang et al. 2009). The sequencing also has lots of new applications using DNA samples, e.g. types of mutations including point mutations, small insertions and deletions, larger insertions, and even chromosomal translocations (Wertheim et al. 2012). Sequencing can also be performed on RNA at both mutation and expression levels. RNA-sequencing (RNA-seq) opens up the transcriptome of non-model organisms and exposes the full complexity of the mammalian transcriptome much of which was hidden from microarrays (McGettigan 2013). RNA-seq can detect RNA editing events and quantify splicing isoforms as well (Buratti et al. 2013). However, the sequencing experiment still faces technical and economic challenges such as data repositories and data analysis (Baker 2012), when compared to a microarray experiment. We still believe that single cell sequencing will displace microarrays in gene expression analysis in the near future.

1.6 Perspectives

Single cell sequencing can be applied in many new insights, e.g. to study the genetically identical cells in the same tissue type which usually be analyzed as a homogenous population which actually not. We can understand how information transmitted from one cell to another (Elowitz et al. 2002). With the new technology, a largely unchartered realm of molecular science will finally be accessible (Pelkmans 2012). We also expect that single cell sequencing can quantitatively display structures, isoforms, and allele-specific expression patterns of both coding genes and non-coding genes (Tang et al. 2011). Furthermore, we hope the technology can integrate genomics with other omic technologies, to provide a more complete understanding of how individual cells and different tissue types are organized and controlled (McGettigan 2013). The single cell sequencing will be applied in many more promising approaches in the near future and is sure to have much more important implications for both basic biological research and medicine as well.

1.7 Opportunities and Challenges

Multiple factors can influence the extent of single-cell heterogeneity within a system, including the source of analyzed tissues (Zhu et al. 2014), biological conditions of the individual cells, or cellular microenvironments. The approaches of single cell transcriptome should avoid, or at least reduce, the variability of clinical samples, in order to define new cell classifications, transitional states, biological distinctions, or biomarkers. For example, the inflammatory stimulus lipopolysaccharide could induce an extensive, previously unobserved, bimodal variation in mRNA abundance and splicing patterns in bone marrow-derived dendritic cells using single-cell RNA-Seq analysis (Shalek et al. 2013). It indicates that stimuli from autocrine-based or external sources may initiate the heterogeneity of the cell, different from our traditional understanding. It is possible that we can identify and discover more sensitive and specific elements responsible for the development of cell heterogeneity as biomarkers to monitor dynamic changes and interactions of gene regulation networks or develop new molecular or cellular therapies. The multiplexed primer design can be introduced for selected targeted genes to lower the pre-amplification primer concentration and primer dimer signals, increase the feasibility of measurements, or make the operation easier for classification of cell type, dissection of heterogeneity, mapping of cellular hierarchy and computational construction of genetic networks (Guo et al. 2013). The challenge in the identification and validation of therapeutic targets is to cell heterogeneity, especially to differ between cancer and pre-cancer cells, activated and non-activated inflammatory cells, or functional and non-functional cells, between sampling micro-regions, between cells at various stages, or between responses to therapy. It is a common challenge for those methodologies including single cell sequencing to define the existence of discovered genes before the occurrence of the disease, before clinical diagnosis, or before the application of therapies. It should be also considered whether the same type of the single cell can be collected for measuring the genomic heterogeneity and frequency of resistant clones in the primary disease before and after a single therapy or multiple manipulations, or during the dynamic follow-up in patients. It is highly expected to have single cell sequencing study with a clear clinical design and definite disease-specific or associated groups, including similar diagnoses, severities, stages, genetic backgrounds, pathogeneses, or therapies. One of the critical parts to apply single-cell sequencing for disease biomarkers and should be to integrate clinical experience and knowledge with bioinformatics from the single cell sequencing and ensure the understanding of diseases. The challenge to translate single cell sequencing analysis from clinical research to clinical application is to standardize the study design and protocol, the study and operation process, data analysis and mining, and the application of clinical bioinformatics. The clinical application of single cell sequencing per se faces the difficulty of problem-solving or question-based study design, and defines the criteria of disease-specific diagnosis, monitoring, or prediction. On the other hand, the single cell sequencing can be used as the tool to screen and identify disease-specific biomarkers which are further developed into a diagnostic panel. It is still difficult to develop the gene panel-based diagnostic kits after the identification, selection, validation, and optimization from the single cell sequencing, due to the unexpected barriers, e.g. for stability and repeatability of the expression, the representative of the targeted panel, or description of the measured results. It is also important to understand the dynamic existence of genomic variation, interaction between genes, and definition of gene scale, origin, rate, and nature.

References

Baker M. Gene data to hit milestone. Nature. 2012;487(7407):282–3.

Blainey PC. The future is now: single-cell genomics of bacteria and archaea. FEMS Microbiol Rev. 2013;37(3):407–27.

Buratti E, Baralle M, Baralle FE. From single splicing events to thousands: the ambiguous step forward in splicing research. Brief Funct Genomic. 2013;12(1):3–12.

Elowitz MB, et al. Stochastic gene expression in a single cell. Science. 2002;297(5584):1183–6.

Gerlinger M, et al. Intratumor heterogeneity and branched evolution revealed by multiregion sequencing. N Engl J Med. 2012;366(10):883–92.

Guo G, Luc S, Marco E, Lin TW, Peng C, Kerenyi MA, Beyaz S, Kim W, Xu J, Das PP, Neff T, Zou K, Yuan GC, Orkin SH. Mapping cellular hierarchy by single-cell analysis of the cell surface repertoire. Cell Stem Cell. 2013;13(4):492–505.

Kohn AB, et al. Single-cell semiconductor sequencing. Methods Mol Biol. 2013;1048:247–84.

Martin J, et al. The impact of next-generation sequencing technology on preimplantation genetic diagnosis and screening. Fertil Steril. 2013;99(4):1054.e3–61.e3.

McGettigan PA. Transcriptomics in the RNA-seq era. Curr Opin Chem Biol. 2013;17(1):4–11.

Metzker ML. Sequencing technologies – the next generation. Nat Rev Genet. 2010;11(1):31–46.

Pelkmans L. Cell biology. Using cell-to-cell variability – a new era in molecular biology. Science. 2012;336(6080):425–6.

Qiu S, et al. Single-neuron RNA-seq: technical feasibility and reproducibility. Front Genet. 2012;3:124.

Ramskold D, et al. Full-length mRNA-seq from single-cell levels of RNA and individual circulating tumor cells. Nat Biotechnol. 2012;30(8):777–82.

Shalek AK, Satija R, Adiconis X, Gertner RS, Gaublomme JT, Raychowdhury R, Schwartz S, Yosef N, Malboeuf C, Lu D, Trombetta JJ, Gennert D, Gnirke A, Goren A, Hacohen N, Levin JZ, Park H, Regev A. Single-cell transcriptomics reveals bimodality in expression and splicing in immune cells. Nature. 2013;498(7453):236–40.

Shapiro E, Biezuner T, Linnarsson S. Single-cell sequencing-based technologies will revolutionize whole-organism science. Nat Rev Genet. 2013;14(9):618–30.

Soon WW, Hariharan M, Snyder MP. High-throughput sequencing for biology and medicine. Mol Syst Biol. 2013;9:640.

Sylvester KG, Longaker MT. Stem cells: review and update. Arch Surg. 2004;139(1):93–9.

Tang F, et al. Deterministic and stochastic allele specific gene expression in single mouse blastomeres. PLoS ONE. 2011;6(6):e21208.

Wang Z, Gerstein M, Snyder M. RNA-seq: a revolutionary tool for transcriptomics. Nat Rev Genet. 2009;10(1):57–63.

Wang J, et al. Genome-wide single-cell analysis of recombination activity and de novo mutation rates in human sperm. Cell. 2012;150(2):402–12.

Wertheim GB, Hexner E, Bagg A. Molecular-based classification of acute myeloid leukemia and its role in directing rational therapy: personalized medicine for profoundly promiscuous proliferations. Mol Diagn Ther. 2012;16(6):357–69.

Wuest SE, Schmid MW, Grossniklaus U. Cell-specific expression profiling of rare cell types as exemplified by its impact on our understanding of female gametophyte development. Curr Opin Plant Biol. 2013;16(1):41–9.

Xue Z, et al. Genetic programs in human and mouse early embryos revealed by single-cell RNA sequencing. Nature. 2013;500(7464):593–7.

Yoshida K, Sanada M, Ogawa S. Deep sequencing in cancer research. Jpn J Clin Oncol. 2013;43(2):110–15.

Zhu Z, Wang DC, Popescu LM, Wang XD. Single-cell transcriptome in the identification of disease biomarkers: opportunities and challenges. J Transl Med. 2014;12:212.

Mengjia Qian Senior Researcher of Zhongshan Hospital Biomedical Research Center, Project leader of Next Generation Sequencing in Fudan University Center for Clinical Bioinformatics. Her main research also includes stem cell isolation and differentiation, telocyte isolation and genomics, and cancer cell epigenetics. She is in charge of the next generation sequencing platform. Her research mainly focuses on the field of immunology. In the recent three years, she has achieved one research fund from the hospital and has published three papers as the first author, including two SCI papers.

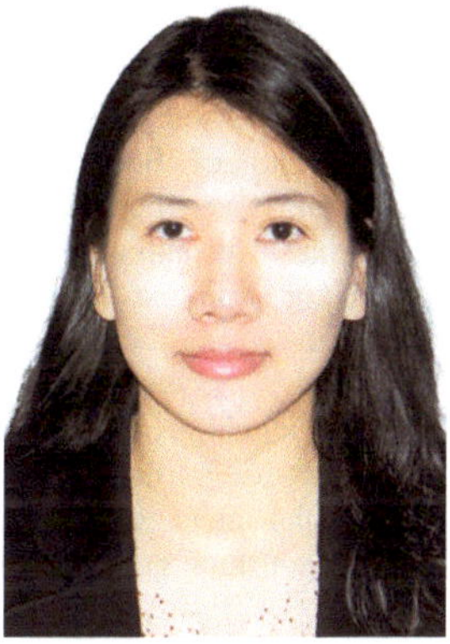

Duojiao Wu Associate Professor of Systems Immunology in the Biomedical Research Center of Fudan University Zhongshan. Principle Investigator and Director of Systems Immunology Group/Platform in Fudan University Center for Clinical Bioinformatics. She was selected and honored as a Shanghai Rising-star Scientist of Science and Technology Commission of Shanghai Municipality, China. Her research focuses on clinical bioinformatics, transplantation immunology, and systems immunology.

Dr. Wu is funded by scientific grants including National Natural Science Foundation. She was invited to serve as reviewers of various international journals and as the faculty speaker in conferences. She has published more than 30 scientific papers in Journal of Translational Medicine, Journal of Proteome Research, and others.

Ding Zhang Dr. Zhang works as a fellow of Respiratory and critical care Medicine at Jiaotong University, Shanghai, China, and further his doctoral programme after Professor Xiangdong, Wang, at department of pulmonary diseases, Zhongshan Hospital, Fudan University, China. His main research is focused on the experimental and clinical research of mechanisms and stem cells therapy combination with telocytes of acute lung injury; he is also interested in lung cancer basic and clinical research and the interaction between lung cancer cells and the immune system; meanwhile, he did some primary single-cell-based lung cancer research as well.

Jian Wang now is studying for a medical master's degree on pulmonary medicine in Zhongshan Hospital, Fudan University. He was graduated from Wenzhou Medical College and had a short-term exchange as visiting student in Medizinische Universität Graz. His work mainly focused on single cell isolation and sequencing.

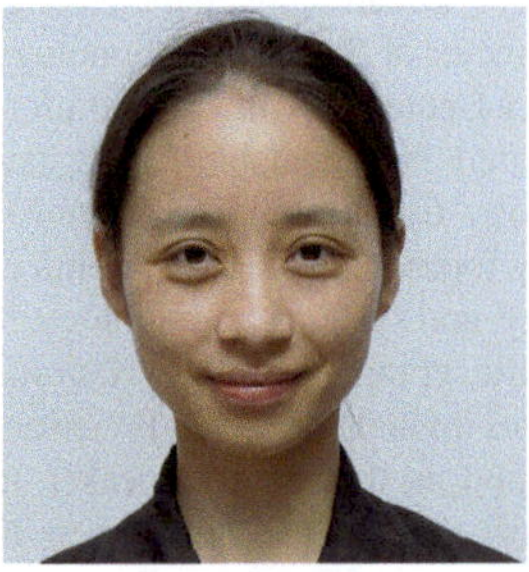

Yunfeng Cheng is a Professor of Hematology Medicine at Fudan University, Deputy Director of Fudan University Center of Clinical Bioinformatics, Deputy Director of Biomedical Research Center, Zhongshan Hospital Fudan University. Director of Department of Hematology, Zhongshan Hospital, Qingpu Branch. Her main research is focused on immune induced blood diseases such as Bone marrow failure, aplastic anemia and immune thrombocytopenia.

She published some famous articles including one on the New England Journal of Medicine (NEJM) as the first author or corresponding author. She got the Pujiang talent grant from Science and Technology Commission of Shanghai Municipality. And she got three grants from National Natural Science foundation of China.

Xiangdong Wang Dr. Wang works as a distinguished professor of Respiratory Medicine at Fudan University, China, adjunct professor of Molecular Bioscience at North Carolina State University, USA, and adjunct professor of Clinical Bioinformatics at Lund University. He serves as a Director of Biomedical Research Center, Fudan University Zhongshan Hospital. He was also pointed as visiting, guest and honor professors in Harbin Medical University (China), Zhejiang University (China), Kyoto Prefectural University of Medicine (Japan), and National Academy of Medicine (Romania). Dr. Wang acts as an Editor-in-Chief of Clinical & Translational Medicine, Journal of Clinical Bioinformatics and Journal of Epithelial Biology & Pharmacology, and Asian Editor of Journal of Cellular & Molecular Medicine. His main research is focused on the role of epithelial cells in the development of organ dysfunction, disease-specific biomarkers, and potential therapies.

Chapter 2
Uncovering Phenotypes with Supercells: Applications to Single-Cell Sequencing

Julián Candia, Jayanth R. Banavar, and Wolfgang Losert

Abstract The so-called "Supercell paradigm" is a method for phenotyping based on single-cell multidimensional data, which has been recently proposed by the authors of this Chapter and collaborators within the larger context of single-cell biology. Supercells are multidimensional objects that represent the collective behavior of groups of cells and carry a distinct phenotype, which is often obscured at the single-cell level due to high cell-to-cell variability. The Supercell framework provides a quantitative assessment of the critical sample size and the number of simultaneous single-cell measurements needed to build a phenotype, which is a key piece of information given the fact that, in many single-cell applications, the number of measured cells and the number of measurements per cell are severely limited due to a variety of constraints, such as experimental costs, technological capabilities, specimen collection procedures, the availability of specialized personnel, and others. In this Chapter, we review the Supercell method and explore the potential for its application to single-cell sequencing datasets.

Keywords Single-cell biology • Single-cell genomics • Cell heterogeneity • Machine learning • Supercells

2.1 Introduction

Heterogeneity from cell to cell is now widely recognized as a key feature of many living systems, which enables their adaptation to changing environmental conditions (Altschuler and Wu 2010). Moreover, similar mechanisms appear to play a significant role in a tumor's ability to survive, proliferate, spread and resist therapy (Marte 2013). Single-cell heterogeneity is often encountered in biomedical research, as well as in the clinical realm, and leads to particular challenges for studies that

J. Candia, Ph.D. (✉)
Center for Human Immunology, National Institutes of Health, 10#7N115,
9000 Rockville Pike, Bethesda, MD 20892, USA
e-mail: julian.candia@nih.gov

J.R. Banavar, Ph.D. • W. Losert, Ph.D.
Department of Physics, University of Maryland, College Park, MD 20742, USA

X. Wang (ed.), *Single Cell Sequencing and Systems Immunology*,
Translational Bioinformatics 5, DOI 10.1007/978-94-017-9753-5_2

are based on a limited number of single cells. One important example is provided by state-of-the-art single-cell genomics technologies, which enable measuring the expression level of all genes in a single cell. However, the number of cells for which all genes can be measured is limited by both cost and instrument capacity. For these new high-dimensional data with limited numbers of data points, data analysis methods that rely on high-dimensional clustering procedures, Gaussian mixture approximations, and other standard classification techniques may be expected to fail. Therefore, it is of paramount importance to address the problem of phenotypic classification when single cells are highly heterogeneous and the number of cells available is small.

Within this context, we have recently proposed the so-called 'Supercell Paradigm' (Candia et al. 2013, 2014) as a general method for single-cell phenotyping that focuses on emergent properties of groups of cells. The key contribution of this method is to provide a quantitative assessment of the critical sample size and number of simultaneous single-cell measurements needed to identify a phenotype with strong predictive power. In (Candia et al. 2013), the Supercell framework was developed and applied to datasets obtained by imaging of cell nuclei and multicolor flow cytometry, as illustrations of the potential of this method to be applied to build multi-parametric phenotypes from different single-cell technologies.

The purpose of this Chapter is to review the Supercell method in detail and to explore the potential for its application in the context of single-cell sequencing data. In order to motivate the need for novel methods of analysis, Sect. 2.2 briefly overviews the challenges arising from high-dimensional single-cell technologies. In Sect. 2.3, we introduce the Supercell paradigm and illustrate the rationale of the method with some examples. In Sect. 2.4, we show the application of the method to single-cell RNA-seq datasets and discuss the potential for further applications. Finally, our Conclusions are stated in Sect. 2.5.

2.2 Phenotypic Heterogeneity and Small-Sample Effects: The Single-Cell Challenge

During the progression from the zygotic stage to adulthood, the aggregate effects of numerous somatic mutations result in the occurrence of several cell lineages with different genotypes in one individual, a phenomenon described as mosaicism (Lupski 2013). Although the true extent of such mosaicism is yet unknown, this phenomenon appears ubiquitous and has led scientists to speculate that each cell in the human body may have a unique genomic signature (Lupski 2013; Shapiro et al. 2013). Many of these mutations are expected to be neutral and others may be disadvantageous and go extinct. Some of them may even be beneficial: For instance, the widespread somatic mutations in the brain, observed in the form of aneuploidy or retro-transposon insertions, might contribute to normal brain function (Baillie et al. 2011; Evrony et al. 2012). However, other somatic mutations are instrumental for the physiologic process of aging (Lopez-Otin et al. 2013) and for the onset of cancer (Tomasetti et al. 2013) and other diseases.

Indeed, single-cell heterogeneity poses challenges as well as huge opportunities in the development and improvement of strategies for the diagnosis and treatment of many diseases (Speicher 2013). For instance, Beckman et al. (2012) have very recently assessed the impact of single-cell heterogeneity, as well as that of genetic instability, in the development of effective nonstandard strategies for personalized cancer treatment. Manifestations of cell heterogeneity in healthy and diseased cell samples have ubiquitously been reported in the growing field of single-cell biology, ranging from human pluripotent embryonic stem cell cultures (de Souza 2012; Tang 2012; Drukker et al. 2012) and apoptosis mechanisms in cancer cell lines (Schmid et al. 2012), to reversible adaptive plasticity in tumors such as human neuroblastoma (Chakrabarti et al. 2012) and pressure-driven shape features of *C. elegans* embryonic cells (Fujita and Onami 2012). For recent reviews of the impact of tumor heterogeneity at different levels (genetic, epigenetic, the tumor microenvironment, the immune response, and other factors such as diet and the microbiota), see Meacham and Morrison (2013), Burrell et al. (2013), Junttila and de Sauvage (2013), and Bedard et al. (2013).

Besides the inherent biological variability from cell to cell, an additional layer of heterogeneity arises from technical noise. Indeed, the accuracy and reliability of single-cell analysis is severely limited by whole-genome and whole-transcriptome amplification noise from a variety of sources (Macaulay and Voet 2014). Although further innovations will be needed to develop the capacity to directly sequence unamplified DNA and RNA derived from single cells, direct library preparation from single-cell genomes has been demonstrated (Falconer et al. 2012; Falconer and Lansdorp 2013) and direct sequencing of single molecules is already a possibility for DNA and RNA (Ozsolak et al. 2009; Coupland et al. 2012).

Another limiting characteristic of current single-cell sequencing studies is the small number of cells investigated, typically in the range from tens to a few hundreds. Within the broader realm of single-cell biology, the inability to have large samples often arises due to technical limitations and cost considerations as well as the nature of the biological/clinical problem at hand. In the field of stem cell research, for instance, stem cells are extremely rare. Thus, identifying and sorting stem cells through flow cytometry yields, even at best, only limited numbers of cells. As an example, long-term hematopoietic stem cells (LT-HSCs) identified via immunophenotypes such as $Lin^{-}Kit^{+}Sca^{+}CD34^{lo}Flt3^{-}$ (Christensen and Weissman 2001) and SLAM (Kiel et al. 2001) represent only about 0.0075 % of the cells from whole bone marrow specimens; thus, more than a million whole bone marrow cells need to be extracted, stained with multiple fluorochromes and sorted in order to yield about one hundred LT-HSCs.

In this Chapter, we will focus on the important case where the number of measured cells is limited for one of the reasons listed above. Furthermore, we will consider situations where cell behavior is so heterogeneous that distinct cell populations have overlapping distributions. Figures 2.1 and 2.2 show schematic

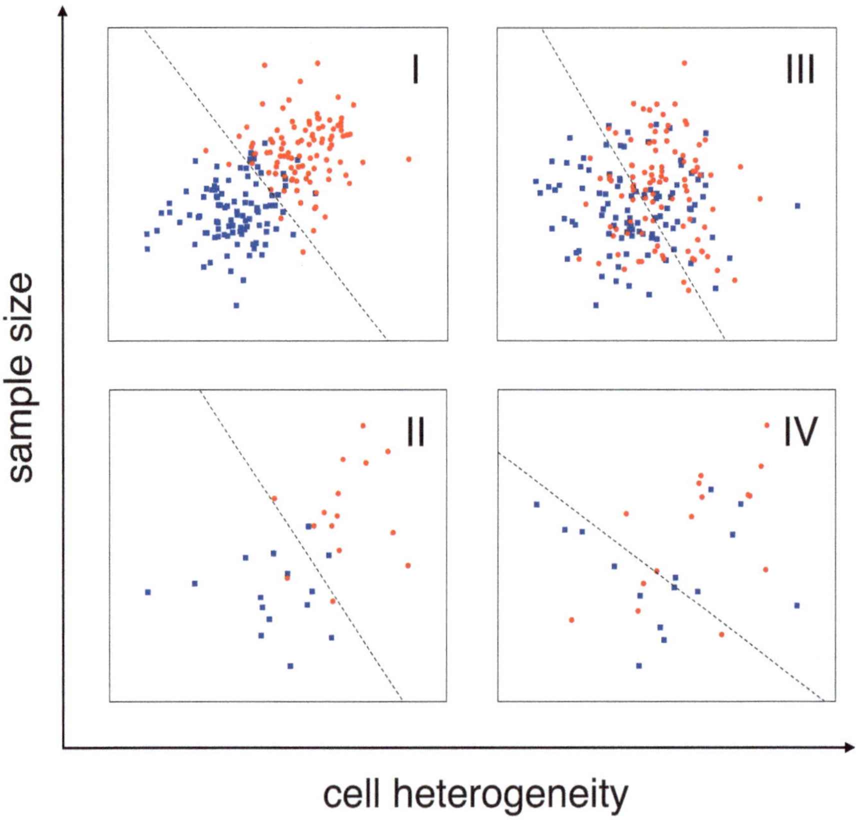

Fig. 2.1 Schematic representation of four scenarios that result from the combinations of low/high cell heterogeneity with small/large sample size. Within each scenario, a *dashed line* shows the linear boundary that optimally separates the two classes (represented by *blue squares* and *red circles*) using a Support Vector Machine (SVM)

representations of the relations between cell heterogeneity, sample size, and the expected classification accuracy of training and test observations. We consider different scenarios in which single-cell measurements are performed on cells that belong to one of two possible classes (i.e. distinct biological phenotypes, such as e.g. cells from a cancer cell line compared with cells from a healthy cell line). On the one hand, cell heterogeneity refers to the observed overlap between the two cell populations, which arise from the biology (depending on how well the chosen biomarkers can inherently distinguish one cell population from another), from the technical procedure (e.g. measurement noise, batch effects, etc.), or more generally from a combination of both biological and technical considerations. On the other hand, sample size refers to the number of single cells measured, which typically depends on considerations of cost, instrument capacity and the number of cells of a particular type. Figure 2.1 schematically considers four main scenarios that

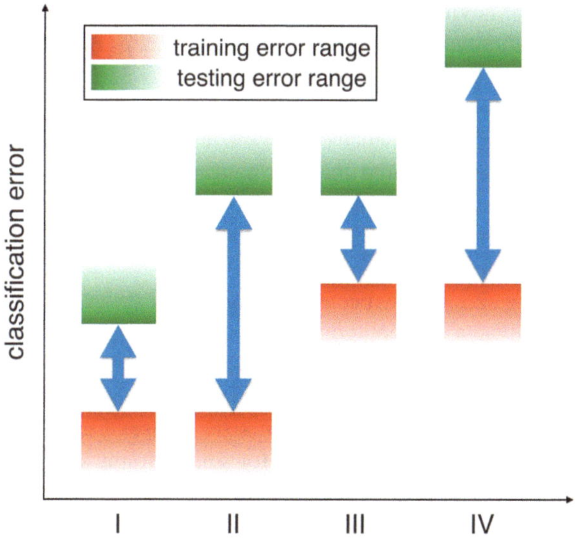

Fig. 2.2 Schematic representation of the expected training (learning) and testing (prediction) classification errors for each of the four scenarios shown in Fig. 2.1. The low-heterogeneity cases (I and II) lead to good class separation and thus small training errors, whereas the high-heterogeneity cases (III and IV) exhibit poorer class separation and larger training errors. On the other hand, large sample sizes (cases I and III) yield a small classification error increase in going from the training to the testing phase, while the increase in more significant with smaller sample sizes (cases II and IV)

result from the combinations of low/high cell heterogeneity with small/large sample size. Within each scenario, a dashed line shows the linear boundary that optimally separates the two classes using a Support Vector Machine (SVM), one popular machine learning method that is widely used as a supervised classifier (more details on SVMs will be given in Sect. 2.3). In supervised classification, the classifier is first built during a so-called *training* or *learning phase*, in which we must know in advance the true classification for each cell. The ability for the classifier to correctly separate the measurement hyperspace in two regions that reflect the true separation between cell classes is quantified by means of the learning error, i.e. the percentage of cells that lie in the wrong side of the classification boundary. Naturally, the learning error is smaller when the cell heterogeneity is smaller, as in Cases I and II, since then it's possible to draw a linear boundary that separates very well the two classes and misclassifies just a few cells. Ranges of expected training errors are schematically depicted in Fig. 2.2 by red-colored fading regions: they are expected to be low in Cases I and II, and higher in Cases III and IV, which are correspondingly characterized by larger cell heterogeneity.

Supervised classifiers are often intended as methods to predict the correct class of new (unknown) instances. This can be used to determine the primary site of a metastatic cancer or to diagnose a disease in a patient from cells obtained with non-invasive or minimally invasive procedures, among many possible applications.

In these examples, it is vital to assess the ability of the method to correctly predict the cell class in the so-called *prediction* or *testing phase*. But, even if we are just interested in inference (that is, to learn the relations between biomarkers to describe phenotypes, rather than to make predictions based on them), it is important to evaluate the classifier's performance during the testing phase to assess whether the classifier has adequately captured the true patterns of classification. Indeed, it is possible to have a classifier that performs well on the training data because it follows the training class labels very closely, but then fails in predicting new instances. This phenomenon is called *overfitting*.

As shown in Fig. 2.2, the expected testing errors given by the green-colored fading regions are always, on average, larger than the corresponding learning errors. However, the increase from learning to testing errors depends on how representative the learning samples are of the true class distributions, which naturally depends on sample size. Thus, we expect large sample sizes (Cases I and III) to yield small training-to-testing error increases, whereas for small sample sizes (Cases II and IV) we expect much larger training-to-testing error increases.

It is interesting to note that, whereas Case I is clearly the best scenario and Case IV is the worst, Cases II and III may yield comparable performance. Yet, there is a delicate balancing act to negotiate the trade-off between effective sample size and effective class heterogeneity in order to find the optimal sweet spot between the two, which yields the optimal performance for a given experimental single-cell dataset. In the next Sections, we develop these ideas further and apply them to both synthetic and true single-cell genomics data.

2.3 The Supercell Paradigm

Highly heterogeneous cell populations (as those represented by Cases III and IV in Fig. 2.1) are not linearly separable and, therefore, the boundary that separates them is ill-defined. In this situation, one solution is to adopt curved decision boundaries that may better fit the separation between classes. To this end, a variety of machine-learning methods such as support vector machines (SVMs) with non-linear kernels, K-nearest neighbors, quadratic and higher-order discriminant analysis, and others, are available to find non-linear class boundaries in high-dimensional measurement space (Hastie et al. 2009; Garteh et al. 2013). However, those methods are also prone to overfitting, a well-known phenomenon in which the classifier performs well on the training data merely because it was trained to follow closely the training instances, but then fails in predicting new instances. As an alternative approach, we build supercell samples as mathematical objects that can be treated in the same way as the directly measured single-cell samples, but which use well-known properties of statistical ensembles to enhance the separation between cell subpopulations. Then, we apply low-variance machine learning methods (such as, e.g., SVM with a linear kernel) on those supercell samples.

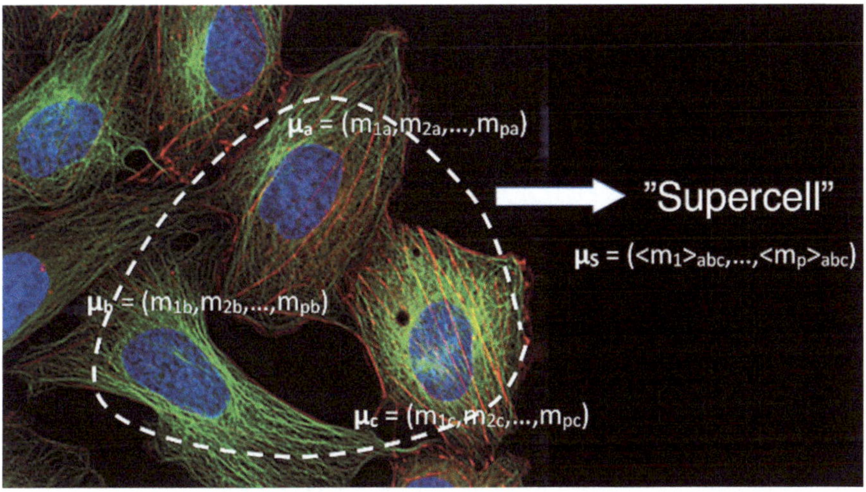

Fig. 2.3 Schematic representation of the supercell averaging procedure. On each cell, p parameters are measured; each cell is represented by a measurement vector μ. A supercell of size N is calculated by randomly selecting N single cells from the sample and then averaging their measurement vectors. By repeating this procedure, we obtain a sample of supercells from the original sample of measured single cells

In order to capture multidimensional cell phenotypes, a 'supercell of size N' is defined as the average of the individual measurement vectors of a group of N randomly chosen cells. By repeatedly taking different random subsets of N cells, 'supercell samples' can be built out of the original single-cell datasets. This procedure is schematically represented in Fig. 2.3, where p parameters are measured on each cell, which is thus quantitatively characterized by a measurement vector μ with p components. In its simplest realization, supercell averaging proceeds by taking N cells at random and averaging their measurement vectors into a supercell vector μ_S. Since the single-cell sample size, N_s, is usually small, supercell averaging is typically performed by selecting cells at random *with replacement*, that is, allowing the same single cell to be chosen more than once. This procedure is indeed similar to the well-known method of *bootstrapping* (Garteh et al. 2013). By iterating this procedure, we obtain a representative sample of N_s' supercells out of the original sample of N_s single cells. Notice that, although the simplest approach builds supercells by combining single cells chosen at random, it is also possible to incorporate additional information to the cell averaging process. In the case of high content multiplexed tissue imaging, for instance, the available 2D or 3D spatial information (localization of each cell, orientation relative to its neighbors, the microenvironment and surrounding extra-cellular matrix, etc.) could be used as inter-cellular level information in the generation of supercells. Analogously, cell cycle phase, cell subtype, etc. may be incorporated to the supercell averaging process for datasets generated by other single-cell technologies such as single-cell genomics.

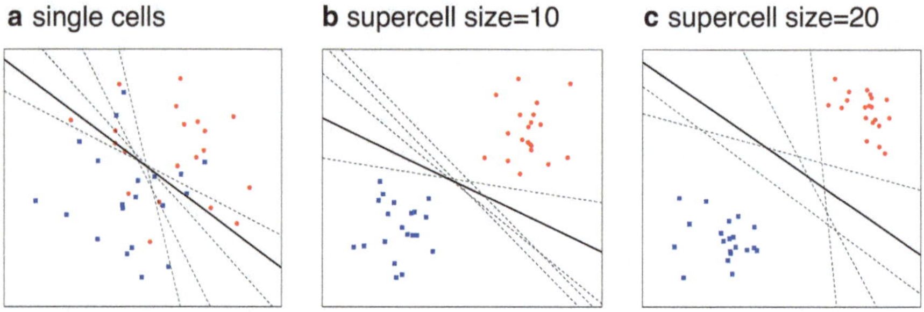

Fig. 2.4 Class separation of 2D synthetic datasets. Samples of 20 cells were obtained for each class, which were randomly generated from uncorrelated 2D normal distributions. The *thick solid line* shows the linear SVM class boundary for the samples displayed (*blue squares* and *red circles*). By generating new samples (not shown), different boundaries are obtained, which are displayed as *thin dashed lines*. (**a**) Learning with single cells, the two populations are not linearly separable. (**b**) Linear separation is achieved by using supercells of size 10. (**c**) By increasing the supercell size to 20, the class separation increases but becomes less robust due to overfitting

After cell averaging, machine learning is used to learn what combination of parameters best distinguishes the different phenotypes. The method implemented in Candia et al. (2013) is a support vector machine (SVM) with a linear kernel, but it can be extended to non-linear mappings that may better reflect the inherent structure of the data. In the linear case, the components of the vector normal to the boundary hyperplane can be straightforwardly interpreted as amplitudes that determine the relative significance of the measured parameters in achieving class separation. Moreover, by introducing appropriate quality functions to balance the tradeoff between separation and robustness, the Supercell paradigm is able to assess the optimal supercell size in order to achieve phenotypic classification when single cells are highly heterogeneous and the number of cells available is small.

Figure 2.4 shows an illustration of the supercell method on 2D synthetic datasets, in which samples of $N_s = 20$ cells were obtained for each class, which were randomly generated from uncorrelated 2D normal distributions. The distributions have the same shape and variance, but their centers are separated. When considering these distributions at the single-cell level, we see that the samples are highly overlapping with no well-defined class boundary (Fig. 2.4a). The thick solid line shows the best class separation obtained from a linear SVM applied to the samples displayed in the figure, where the two classes are represented by blue squares and red circles, respectively. In order to represent the fluctuations arising from the combination of cell population overlap and small sample size, the dashed lines show SVM class boundaries obtained by generating new samples of 20 cells each (these additional samples are not displayed in the figure, only the resulting class boundaries are). By generating supercell samples, the cell populations separate, as expected on the basis of the central limit theorem. For simplicity, we choose a supercell sample size, N_s', equal to the original single-cell sample size $N_s = 20$. Figure 2.4b shows the linear separation achieved by using supercell size $N = 10$. As before, fluctuations arising from different samples are shown by dashed lines. In Fig. 2.4c, the supercell size has been increased to $N = 20$. Correspondingly, the class separation is even larger, at the expense of a larger variation in the orientation of the class boundaries as different cell samples are considered. This is a manifestation of overfitting, since in this case the supercells used to learn the class boundary are highly correlated with each other and very close to the overall class means.

It should be noted that the orientation of the class boundary is affected by different factors. One of them is single-cell sampling: from each class distribution, only N_s single cells are actually measured. When N_s is relatively small, strong sampling effects are to be expected and thus, large boundary fluctuations will arise from considering different sets of single-cell ensembles. On the other hand, when generating supercell samples out of the measured single cells, there are many different ways of choosing N single cells to average together into each supercell, which leads to an additional source for class boundary fluctuations. Finally, it is important to notice that the supercell averaging procedure has the combined effect of shrinking the cell distribution in parameter space, as well as that of modifying the shape of the distribution closer to a normal distribution, which is the expected effect of averaging due to the central limit theorem. Thus, if the original single-cell distribution is skewed, fat-tailed, or has many outlier observations beyond the expected range of a normal distribution, the shape of the resulting supercell distribution will be significantly different from the corresponding single-cell one.

The orientation of the class boundary conveys important information about the relative importance of the different measures. Indeed, when the number of measured parameters per cell, p, is very large, the orientation of the class boundary allows us to rank-order the measures, remove the least significant one, and reiterate the learning procedure, a process known as *recursive feature elimination* (Guyon et al. 2006). Thus, based on the considerations mentioned above, it is important to stress the fact that supercell averaging allows us to optimally characterize cell phenotypes based on class labeling and, thus, to work around the difficulties imposed by cell heterogeneity within each class.

These phenotypes are collective properties of cells within each class and do not necessarily reflect the best combination of parameters to characterize single cells within each class.

In Candia et al. (2013), we have developed and applied the Supercell/SVM paradigm to datasets obtained by different single-cell technologies, e.g. imaging of cell nuclei and multicolor flow cytometry. As a case example of the latter, we focused on the challenging problem of building molecular phenotypes to characterize the differences between two non-infectious uveitides (the ocular manifestations of sarcoidosis and Behçet's disease), which are very difficult to diagnose in the clinic and require different treatments. By performing two scattering and 14 fluorescent measurements on each cell, samples from 7 sarcoidosis and 6 Behçet's patients were measured. Since the cohort was small, prediction testing was carried out by a jackknife (leave-one-out) cross-validation procedure. The SVM boundary allows one to rank-order the 16 measures from most to least significant, according to the components of the vector normal to the hyperplane that separates the two diseases. Thus, one can selectively remove the least significant measurements from the list

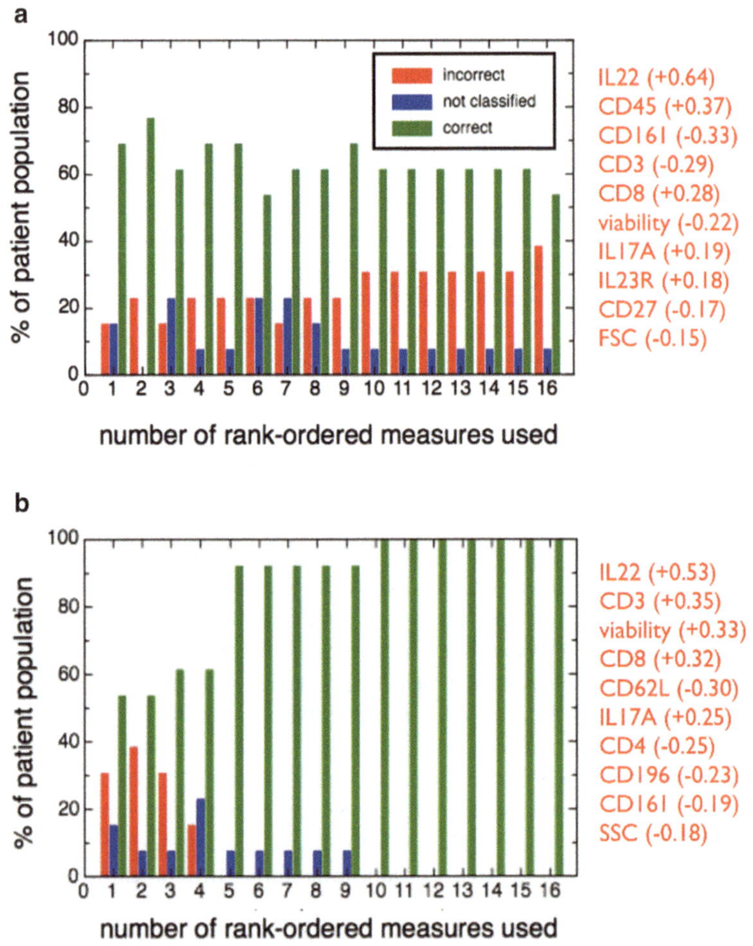

Fig. 2.5 Example of supercell phenotyping using multicolor flow cytometry. Leave-one-out (jackknife) cross-validation results for sarcoidosis versus Behçet's disease using supercells of size N = 500, where each patient is represented by a cloud of 100 supercells, as a function of the number of rank-ordered measures used: (**a**) All cells; (**b**) CD8+ T cells. The *bars* show percentages of correct (*green*), unclassified (*blue*) and incorrect (*red*) predictions. To the right of each panel, the list of the top 10 rank-ordered measures is shown (Adapted from Candia et al. 2013)

and explore the minimal number of measures needed to correctly predict the class of all (or at least most of) the samples. Figure 2.5 shows jackknife results for supercells of size N = 500, where each patient is represented by a cloud of 100 supercells, as a function of the number of rank-ordered measures used. The list of the top 10 measures is shown to the right of each panel. Figure 2.5a shows jackknife results based on all cells, while Fig. 2.5b displays results based on $CD8^+$ T cells, a subpopulation that can be determined by manual gating ($CD3^+$ $viab^-$ $CD8^+$ $CD4^-$) and typically represents about 5 % of the peripheral blood sample.

Since each patient is represented by a cloud of supercells, a prediction was made only when more than 95 % of those supercells lie on any one side of the SVM boundary. Correct predictions are shown by green bars, incorrect predictions by red bars, while unclassified samples are shown in blue. While predictions based on all cells are very poor, for $CD8^+$ T cells no failed predictions are incurred when five or more measures are used. Therefore, the top five measures listed in Fig. 2.5b, can be linearly combined in order to be used on $CD8^+$ T cells as molecular phenotypes that distinguish the two diseases.

2.4 Applications of the Supercell Framework to Single-Cell Sequencing: A Case Study

In this Section, we will work out a case study using publicly available single-cell RNA-seq datasets. The purpose of this Section is to illustrate possible applications of the ideas discussed earlier to the kinds of datasets produced by state-of-the-art single-cell sequencing technologies. Our main focus here is not on the biology, but rather, on the method and its potential as an analysis tool on datasets characterized by highly-overlapping phenotypes obtained from highly-dimensional single-cell datasets that span a limited number of cells, typically around or below 100 cells.

To this end, we will use publicly available data (http://www.ncbi.nlm.nih.gov/geo/query/acc.cgi?acc=GSE52583) that correspond to microfluidic single-cell RNA-seq on 198 individual mouse lung epithelial cells at four different stages through development, namely E14.5, E16.5, E18.5, and AT2 (adult). In order to control background and normalization, these datasets include 92 external RNA (ERCC) spike-ins; moreover, one no-cell and two 200-cell bulk control samples were generated for time point E18.5 (see http://www.ncbi.nlm.nih.gov/geo/query/acc.cgi?acc=GSE52583; Treutlein et al. 2014 for further technical details). In our analysis, we used 45 E14.5 cells, 27 E16.5 cells, 34 E18.5 cells (which correspond to replicate # 2 in http://www.ncbi.nlm.nih.gov/geo/query/acc.cgi?acc=GSE52583) and 46 AT2 (adult) cells, totaling 152 cells. Based on the datasets in (http://www.ncbi.nlm.nih.gov/geo/query/acc.cgi?acc=GSE52583), Treutlein et al. (2014) have very recently confirmed the basic outlines of the conventional model of cell type diversity in the distal lung, as well as discovering a large number of novel transcriptional regulators and cell type markers that discriminate between different cell populations.

After obtaining RNA-seq expression values in terms of Fragments Per Kilobase of transcript per Million mapped reads (FPKM), we transformed them to $\log_2(\text{FPKM} + 0.5)$. The RNA-seq matrix is typically very sparse, with most entries corresponding to zero transcripts (which, in our log-transformed scale, are represented by -1 values). Thus, single-cell genomics poses particular challenges for data analysis due to low signal-to-noise ratios, small sample sizes, and with the additional complication arising from their attributes spanning an extremely high-dimensional space.

The issue of high-dimensional spaces can be addressed by implementing one of several possible feature selection schemes (Guyon et al. 2006). In Sect. 2.3, we showed one example in which a panel of multiple flow cytometry measures was first used to achieve the separation of phenotypes, then followed by *recursive feature elimination* (also called *backward stepwise selection*) to sequentially remove the least important markers in a top-down fashion until a core set of markers was found. Alternatively, one could start by considering individual measures and selecting the one that has the smallest classification error, and iterating in a forward, stepwise manner, adding more predictors to the model, one at a time. This bottom-up procedure is called *recursive feature addition* or *forward stepwise selection*. Other approaches to feature selection are hybrid implementations that add predictors sequentially, in analogy to forward selection, but at each step may also remove measures that no longer provide an improvement in the model classification.

In this Section, however, we use a different approach. We focus on biologically meaningful gene sets that rely on relevant pathway information. This approach is complementary to fully unbiased methods of gene set selection that are agnostic to the biology. Indeed, whereas the former approach may lead to more focused knowledge in the context of specific biological processes, the latter may lead to the discovery of new molecular mechanisms and thus open up new avenues of research. Recalling that the RNA-seq data we are concerned with here correspond to individual mouse lung epithelial cells at different developmental stages, we will focus on a set of genes that have roles as oncogenes or tumor suppressor genes in KEGG pathways associated with small cell lung cancer (http://www.kegg.jp/kegg-bin/show_pathway?mmu05222) and non-small cell lung cancer (http://www.kegg.jp/kegg-bin/show_pathway?mmu05223). After disregarding two genes (Alk and Cdkn2a) that are uniformly undetected in all the 152 single-cell samples considered here, we are left with a panel of nine target genes, namely: Eml4, Fhit, Kras, Myc, Pten, Rarb, Rassf1, Rb1, and Trp53.

The SVM approach used in the previous Section dealt with a two-class learning problem based on multidimensional single-cell measurements in a nearly continuous range. The example discussed in more detail (Fig. 2.5) was based on multicolor flow cytometry intensity measurements, which typically lie above detection thresholds. In contrast, RNA-seq data are characterized by sparse expression matrices with many zeros and the learning problem we are considering here has four classes with a natural progression given by cell developmental stage. Rather than using SVM boundaries, this kind of classification problem is better solved by random forests, a generalization of decision trees in which instances are classified depending on a sequence of binary decisions based on measurement thresholds (Garteh et al. 2013; Breiman 2001). A variety of random forest algorithms has been successfully applied to many applications in bioinformatics (see e.g. Strobl et al. 2007 and references therein). Random forests can be applied to a wide range of prediction problems, even if they are nonlinear and involve complex high-order interaction effects, and they produce variable importance measures for each predictor variable.

Figure 2.6 shows results of random forests applied to the single-cell datasets described above. Based on so-called *out-of-bag* (OOB) data (see Garteh et al.

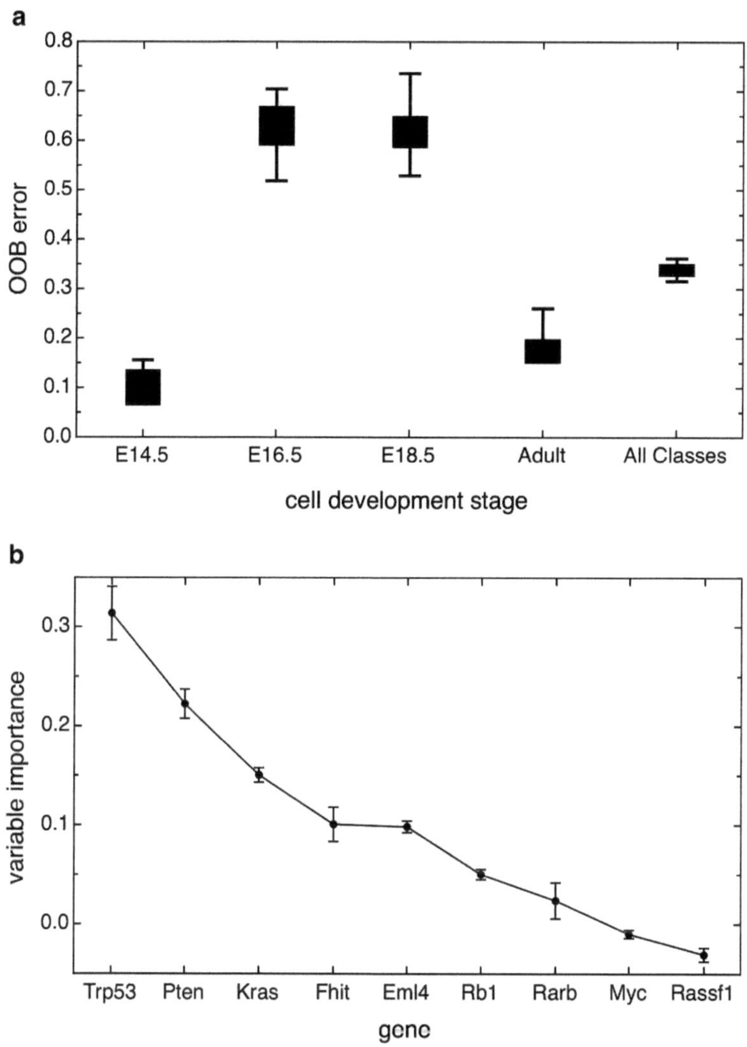

Fig. 2.6 Random forest results for RNA-seq individual mouse lung epithelial cells in 4 different stages of development: E14.5, E16.5, E18.5 and AT2 (adult). We focus on a set of 9 genes involved in lung cancer pathways. (**a**) *Box plots* showing so-called out-of-bag (OOB) error rate distributions (calculated from over 100 iterations of random forests built using 1,000 trees each) for each class. The overall mean OOB error rate for single-cell classification is 34 %. (**b**) Variable importance of each of the genes in the gene panel computed as the mean decrease of accuracy. Variable importance values have been normalized to add up to 1. Here, larger values indicate increased importance for the classification decision. The span of the vertical bars represents one standard deviation above and below the mean variable importance

2013; Breiman 2001 for background information and http://cran.r-project.org/web/packages/randomForest/index.htmlfor method implementation details), Fig. 2.6a shows the OOB error rates for the classification of cells in each of the four classes based on RNA-seq expression of genes in the selected gene panel. The box plot displays the distribution of OOB error rates over 100 iterations of random forests built using 1,000 trees each. We observe that, whereas the phenotypes for the early development stage E14.5 and the adult one appear well characterized (with mean OOB error rates of 11 % and 19 %, respectively), the intermediate E16.5 and E18.5 stages are poorly characterized (with mean error rates above 60 %). The overall mean OOB error rate for single-cell classification is 34 %. By recording the mean decrease of accuracy of predictions in the out-of-bag samples when a given predictor (gene) is excluded, we obtain a measure of the so-called *variable importance* for that predictor. Figure 2.6b shows the variable importance of each of the genes in the gene panel for random forest learning based on single cells. Variable importance values have been normalized to add up to 1. Notice that larger values indicate increased importance for the classification decision.

Now we can incorporate supercell averaging: following the rationale described in Sect. 2.3, we can generate a supercell ensemble of 45 E14.5 supercells, 27 E16.5 supercells, 34 E18.5 supercells and 46 AT2 (adult) supercells. As before, one supercell of size N is obtained by averaging the single-cell measurement vectors (in this case, associated with the expression of multiple genes) over N randomly chosen single cells with replacement (i.e. allowing the same cell to be chosen more than once). Based on one such supercell ensemble, we apply the random forest learning method using 1,000 trees. Then, we iterate this procedure 100 times and measure OOB error rate distributions, as we did on the (original) single cell datasets.

Figure 2.7a shows the mean OOB error resulting from random forests as a function of supercell size. As expected due to supercell averaging shrinkage via the central limit theorem, the mean OOB error decreases monotonically with supercell size. In order to choose the optimal supercell size, we need to implement a criterion to choose the appropriate degree of flexibility of our model. In other words, we need to optimize the so-called *bias-variance tradeoff*: as we average using supercells, the distributions shrink and their overlap decreases, making it easier to identify different classes; however, this decreased bias comes at the expense of an increased variance due to the introduction of effective correlations between supercells (i.e. supercell learning instances are not truly statistically-independent observations, as single cells are, and the equivalent sample size of effectively independent supercell observations becomes smaller than the original single-cell sample size). As a simple approach to adjust the OOB error rate to account for the model size (i.e. the choice of supercell size), we roughly estimate the prediction rate using an ad-hoc Information Criterion of the form $\mathrm{IC} = \mathrm{OOB\ error} + d\ \sigma\ \sqrt{N}$, where the second term is a penalty from using a high-dimensional parameter space of dimension d and supercells of size N. The $\sqrt{N}$ dependence stems from the fact that, due to the central limit theorem, the width of supercell distributions shrinks as $\sqrt{N}$, while σ represents an estimate of the overall variance of the error ε associated with each response measurement. In this case, we adopt $\sigma = 0.01$, while the dimensionality is $d = 9$. Figure 2.7b shows IC as

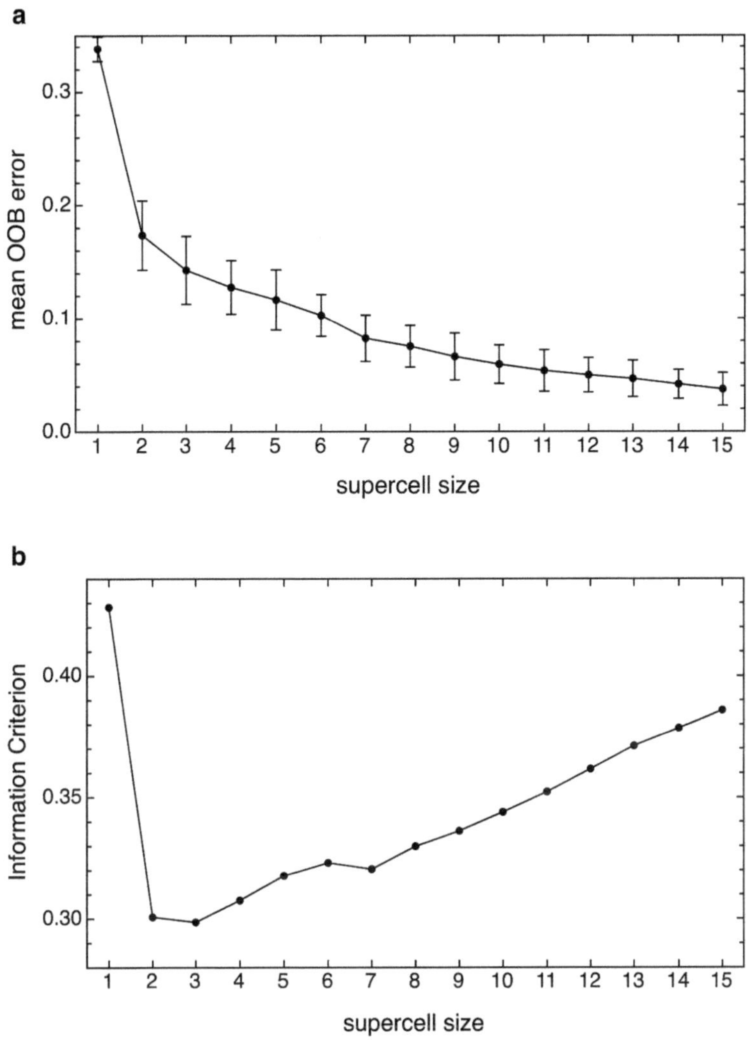

Fig. 2.7 Random forest results as a function of supercell size, using 1,000 trees in each random forest and averaging over 100 supercell realizations. (**a**) The mean OOB error decreases monotonically with supercell size, as expected due to supercell averaging shrinkage via the central limit theorem. (**b**) By considering an ad-hoc Information Criterion to balance the bias-variance tradeoff, the optimal supercell size is $N = 3$ (See text for details)

a function of supercell size. After adjusting by IC, we observe that the estimate for the prediction error is a minimum for $N = 3$; for $N > 3$, the penalty due to increased cell averaging overrides the gains obtained due to smaller OOB error rates, thus leading to increasing IC values. It should be pointed out that, with larger datasets, the validation set and cross-validation methods may be implemented to directly estimate prediction errors.

Finally, Fig. 2.8 shows random forest results for supercells of size N = 3. Figure 2.8a shows OOB error rate distributions (over 100 iterations of random forests built using 1,000 trees each) for each class. By comparing them to the single-cell OOB error rate distributions in Fig. 2.6a, we observe that, at the expense of small increases in the error rates of classes E14.5 and AT2, vast improvements in the classification of the intermediate development stages (E16.5 and E18.5) are achieved. The overall mean OOB error rate is also very significantly reduced. Figure 2.8b shows the importance of each of the genes in the gene panel computed as the mean decrease of accuracy. As before, variable importance values have been normalized to add up to 1 and larger values indicate increased importance for the classification decision. Standard deviations (shown by vertical bars) are larger in the supercell case compared with the single-cell case, as expected from the fact that, in the former, we average over different decision trees in the random forest as well as over different supercell realizations. The assessed relative importance of the various genes in the panel, however, does not display any significant differences.

2.5 Conclusions

The Supercell paradigm is a method for phenotyping based on single-cell multidimensional data, which has been recently proposed by the authors of this Chapter and collaborators within the larger context of single-cell biology, focusing on applications to multicolor flow cytometry and high-content image-based phenotyping (Candia et al. 2013). Supercells are multidimensional objects that represent the collective behavior of groups of cells; within this approach, supercells represent the building blocks of healthy and diseased phenotypes. From a conceptual standpoint, this approach naturally incorporates emergent behavior and thus cell heterogeneity, usually regarded as a roadblock in the pursuit of characterizing single-cell-level behavior, becomes the fundamental conceptual unit to identify collective phenotypes. From a practical perspective, the Supercell framework provides a quantitative assessment of the critical sample size and the number of simultaneous single-cell measurements needed to build a phenotype, which is a key piece of information given the fact that, in many single-cell applications, the number of measured cells and the number of measurements per cell are severely limited due to a variety of constraints, such as experimental costs, technological capabilities, specimen collection procedures, the availability of specialized personnel, and others.

Single-cell sequencing technologies generate datasets that pose particular challenges for data analysis due to low signal-to-noise ratios, small sample sizes, and extremely high-dimensional predictor spaces. In this Chapter, we discussed ways in which supercells could provide useful conceptual and computational means to deal with some of those challenges. Hopefully, these tools and ideas will stimulate further work and will contribute to advance the emerging and very promising field of single-cell biology.

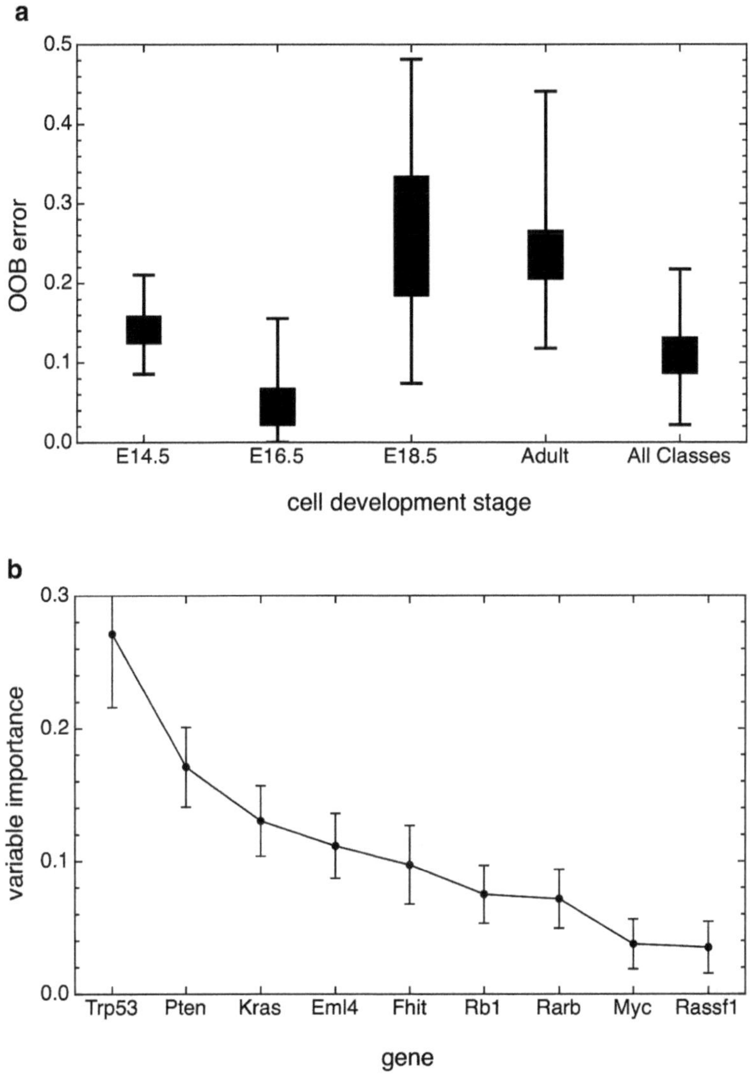

Fig. 2.8 Random forest results for supercells of size N = 3. (**a**) *Box plots* showing OOB error rate distributions (over 100 iterations of random forests built using 1,000 trees each) for each class. The overall mean OOB error rate is 14 %. (**b**) Variable importance of each of the genes in the gene panel computed as the mean decrease of accuracy. Variable importance values have been normalized to add up to 1. Here, larger values indicate increased importance for the classification decision. The span of the vertical bars represents one standard deviation above and below the mean variable importance

Acknowledgments We acknowledge our coauthors A. Biancotto, K. Cao, P. Dagur, M. Driscoll, A. Maritan, R. Maunu, J. P. McCoy Jr., R. B. Nussenblatt, H. N. Sen, and L. Wei, whose contributions to the Supercell approach (Candia et al. 2013) are extensively described in this Chapter. J. C. was supported by NIH Award Number T32CA154274 from the National Cancer Institute. The content is solely the responsibility of the authors and does not necessarily represent the official views of the National Cancer Institute or the National Institutes of Health.

References

Altschuler S, Wu LF. Cellular heterogeneity: do differences make a difference? Cell. 2010;141:559.

Baillie JK, Barnett MW, Upton KR, Gerhardt DJ, Richmond TA, et al. Somatic retrotransposition alters the genetic landscape of the human brain. Nature. 2011;479:534–7.

Beckman RA, Schemmann GS, Yeang C-H. Impact of genetic dynamics and single-cell heterogeneity on development of nonstandard personalized medicine strategies for cancer. Proc Natl Acad Sci U S A. 2012;109(36):14586–91.

Bedard PL, Hansen AR, Ratain MJ, Siu LL. Tumour heterogeneity in the clinic. Nature. 2013;501(7467):355–64.

Breiman L. Random forests. Mach Learn. 2001;45(1):5–32.

Burrell RA, McGranahan N, Bartek J, Swanton C. The causes and consequences of genetic heterogeneity in cancer evolution. Nature. 2013;501(7467):338–45.

Candia J, Maunu R, Driscoll M, Biancotto A, Dagur P, McCoy Jr JP, Sen HN, Wei L, Maritan A, Cao K, Nussenblatt RB, Banavar JR, Losert W. From cellular characteristics to disease diagnosis: uncovering phenotypes with supercells. PLoS Comput Biol. 2013;9:e1003215.

Candia J, Banavar JR, Losert W. Understanding health and disease with multidimensional single-cell methods. J Phys Condens Matter. 2014;26:073102.

Chakrabarti L, Abou-Antoun T, Vukmanovic S, Sandler AD. Front Oncol. 2012;2:82.

Christensen JL, Weissman IL. Flk-2 is a marker in hematopoietic stem cell differentiation: a simple method to isolate long-term stem cells. Proc Natl Acad Sci U S A. 2001;98(25):14541–6.

Coupland P, Chandra T, Quail M, Reik W, Swerdlow H. Direct sequencing of small genomes on the Pacific Biosciences RS without library preparation. Biotechniques. 2012;53:365–72.

de Souza N. Taming stem cell heterogeneity. Nat Method. 2012;9(7):645.

Drukker M, Tang C, Ardehali R, Rinkevich Y, Seita J, Lee AS, Mosley AR, Weissman IL, Soen Y. Isolation of primitive endoderm, mesoderm, vascular endothelial and trophoblast progenitors from human pluripotent stem cells. Nat Biotechnol. 2012;30(6):531–42.

Evrony GD, Cai X, Lee E, Hills LB, Elhosary PC, et al. Single-neuron sequencing analysis of L1 retrotransposition and somatic mutation in the human brain. Cell. 2012;151:483–96.

Falconer E, Lansdorp PM. Strand-seq: a unifying tool for studies of chromosome segregation. Semin Cell Dev Biol. 2013;24:643–52.

Falconer E, Hills M, Naumann U, Poon SS, Chavez EA, et al. DNA template strand sequencing of single-cells maps genomic rearrangements at high resolution. Nat Methods. 2012;9:1107–12.

Fujita M, Onami S. Cell-to-cell heterogeneity in cortical tension specifies curvature of contact surfaces in Caenorhabditis elegans embryos. PLoS ONE. 2012;7:e30224.

Garteh J, Witten D, Hastie T, Tibshirani R. An introduction to statistical learning. New York: Springer; 2013.

Guyon I, Gunn S, Nikravesh M, Zadeh L, editors. Feature extraction: foundations and applications. New York: Springer; 2006.

Hastie T, Tibshirani R, Friedman J. The elements of statistical learning. 2nd ed. New York: Springer; 2009.
http://cran.r-project.org/web/packages/randomForest/index.html
http://www.kegg.jp/kegg-bin/show_pathway?mmu05222
http://www.kegg.jp/kegg-bin/show_pathway?mmu05223
http://www.ncbi.nlm.nih.gov/geo/query/acc.cgi?acc=GSE52583
Junttila MR, de Sauvage FJ. Influence of tumour micro-environment heterogeneity on therapeutic response. Nature. 2013;501(7467):346–54.
Kiel MJ, Yilmaz ÖH, Iwashita T, Yilmaz OH, Terhorst C, Morrison SJ. Cell. 2001;121(7):1109–21.
Lopez-Otin C, Blasco MA, Partridge L, Serrano M, Kroemer G. The hallmarks of aging. Cell. 2013;153:1194–217.
Lupski JR. Genetics. Genome mosaicism – one human, multiple genomes. Science. 2013;341: 358–9.
Macaulay IC, Voet T. Single cell genomics: advances and future perspectives. PLoS Genet. 2014;10(1):e1004126.
Marte B (ed) Tumour heterogeneity. Nature. 2013;501(67):327–72.
Meacham CE, Morrison SJ. Tumour heterogeneity and cancer cell plasticity. Nature. 2013;501(7467):328–37.
Ozsolak F, Platt AR, Jones DR, Reifenberger JG, Sass LE, et al. Direct RNA sequencing. Nature. 2009;461:814–18.
Schmid J, Dussmann H, Boukes GJ, Flanagan L, Lindner AU, O'Connor CL, Rehm M, Prehn JH, Huber HJ. J Biol Chem. 2012;287(49):41546–59.
Shapiro E, Biezuner T, Linnarsson S. Single-cell sequencing-based technologies will revolutionize whole-organism science. Nat Rev Genet. 2013;14(9):618–30.
Speicher MR. Single-cell analysis: toward the clinic. Genome Med. 2013;5:74.
Strobl C, Boulesteix A-L, Zeileis A, Hothorn T. Bias in random forest variable importance measures: illustrations, sources and a solution. BMC Bioinformatics. 2007;8:25.
Tang DG. Understanding cancer stem cell heterogeneity and plasticity. Cell Res. 2012;22(3):457–72.
Tomasetti C, Vogelstein B, Parmigiani G. Half or more of the somatic mutations in cancers of self-renewing tissues originate prior to tumor initiation. Proc Natl Acad Sci U S A. 2013;110:1999–2004.
Treutlein B, Brownfield DG, Wu AR, Neff NF, et al. Reconstructing lineage hierarchies of the distal lung epithelium using single-cell RNA-seq. Nature. 2014;509(7500):371–5.

Julián Candia is a Staff Scientist with the Center of Human Immunology at the National Institutes of Health (Bethesda, MD). He earned a degree as 'Licenciado en Física' from the University of La Plata (Argentina) in 1999, and his Ph.D. in Physics from the same university in 2004. Dr. Candia's research interests are focused on analyzing large-scale biological datasets with a variety of computational and analytical techniques at the crossroads of physics, statistics, and computer science. His particular interests are in integrating multiple types of biomedical information, including cell-biological and biophysical information, e.g. gene and microRNA expression, cell shapes, and immunophenotypes, with the goal of contributing innovative ideas to the most pressing problems in current biomedical research. For more information, see *juliancandia.com*.

Jayanth R. Banavar is Dean of the College of Computer, Mathematical, and Natural Sciences at the University of Maryland. Prior to his current appointment, Dr. Banavar served as Distinguished Professor and George A. and Margaret M. Downsbrough Department Head of Physics at Pennsylvania State University. He received a Bachelor of Science with honors and a Master of Science in physics from Bangalore University. He earned his Ph.D. in physics from the University of Pittsburgh. A fellow of the American Physical Society and the American Association for the Advancement of Science, he has more than 250 publications in refereed journals, 11 book chapters, a book he co-edited, and three patents. Much of Dr. Banavar's recent work has applied the techniques of statistical physics to solve interdisciplinary problems, explaining, for example, why biological molecules tend to curl up into helices, or to explain why coral reefs support such a rich biodiversity. Frequently, the goal has been to identify an underlying mathematical principle to provide an elegant explanation of natural phenomena.

Wolfgang Losert is Professor of Physics and Associate Dean for Research at the University of Maryland and co-founder and Director of the University of Maryland-National Cancer Institute Partnership for Cancer Technology. He received a Diploma in Applied Physics in 1995 from the Technical University of Munich, and a Doctorate in Physics from the City College of the City University of New York in 1998.

Dr. Losert's research group studies the dynamics of living systems, with a focus on new approaches to measure and understand the spatial patterns and dynamics of cells and tissues. A particular aim of the work is to better understand the physical characteristics of cancer, such as the shape, dynamics and heterogeneity of cancer cells. Through his research, Dr. Losert contributes novel approaches to harness the emerging abundance of quantitative data in biological and medical research.

Chapter 3
Single-Cell Sequencing of Microorganisms

Ying Chen and Fengping Wang

Abstract Only a small fraction (estimated to be less than 1 %) of microbial species on Earth can be cultivated in the laboratory; thus, the standard microbial research methods based on pure culture isolation and observation can provide only very limited information about an environmental microbial community. The development and successful application of microbial small sub-unit ribosomal RNA (16S rRNA) gene PCR analysis has greatly expanded our knowledge of the diversity and phylogeny of microorganisms. Novel, yet-uncultivated microorganisms have been continually discovered by the 16S rRNA gene approach, revealing an "uncultured microbial majority", which is estimated to comprise 40–50 as yet-uncultivated candidate phyla of bacteria and a similar number of as-yet uncultivated major lineages of archaea (Rappé and Giovannoni 2003). In an allusion to astrophysics that highlights its importance, the "uncultured microbial majority" has been called "biological dark matter" or "microbial dark matter" (Marcy et al. 2007). Recent achievements in metagenomics (genomic sequences from the entire environmental community) and single-cell genomics are now opening the window to observation and analysis of this "biological dark matter".

Keywords Uncultivated • Dark matter • Single cell • Whole genome amplification • Genome

Y. Chen
State Key Laboratory of Microbial Metabolism, School of Life Sciences and Biotechnology, Shanghai Jiao Tong University, 800 Dongchuan Road, Minghang, 200240 Shanghai, China

F. Wang, Ph.D. (✉)
State Key Laboratory of Microbial Metabolism, School of Life Sciences and Biotechnology, Shanghai Jiao Tong University, 800 Dongchuan Road, Minghang, 200240 Shanghai, China

State Key Laboratory of Ocean Engineering, Shanghai Jiao Tong University, 200240 Shanghai, China
e-mail: fengpingw@sjtu.edu.cn

X. Wang (ed.), *Single Cell Sequencing and Systems Immunology*,
Translational Bioinformatics 5, DOI 10.1007/978-94-017-9753-5_3

3.1 Introduction

Only a small fraction (estimated to be less than 1 %) of microbial species on Earth can be cultivated in the laboratory; thus, the standard microbial research methods based on pure culture isolation and observation can provide only very limited information about an environmental microbial community. The development and successful application of microbial small sub-unit ribosomal RNA (16S rRNA) gene PCR analysis has greatly expanded our knowledge of the diversity and phylogeny of microorganisms. Novel, yet-uncultivated microorganisms have been continually discovered by the 16S rRNA gene approach, revealing an "uncultured microbial majority", which is estimated to comprise 40–50 as yet-uncultivated candidate phyla of bacteria and a similar number of as-yet uncultivated major lineages of archaea (Rappé and Giovannoni 2003). In an allusion to astrophysics that highlights its importance, the "uncultured microbial majority" has been called "biological dark matter" or "microbial dark matter" (Marcy et al. 2007). Recent achievements in metagenomics (genomic sequences from the entire environmental community) and single-cell genomics are now opening the window to observation and analysis of this "biological dark matter".

Single-cell sequencing analyzes the genomic information of individual cells with the aid of rapidly advancing sequencing methodology. This technology provides the genomic information of an individual cell within its microenvironment. Generally, single-cell sequencing is comprised of two parts: single cell isolation and whole genome amplification (Fig. 3.1). Here we provide a short review of the methodology of single-cell sequencing and its recent applications.

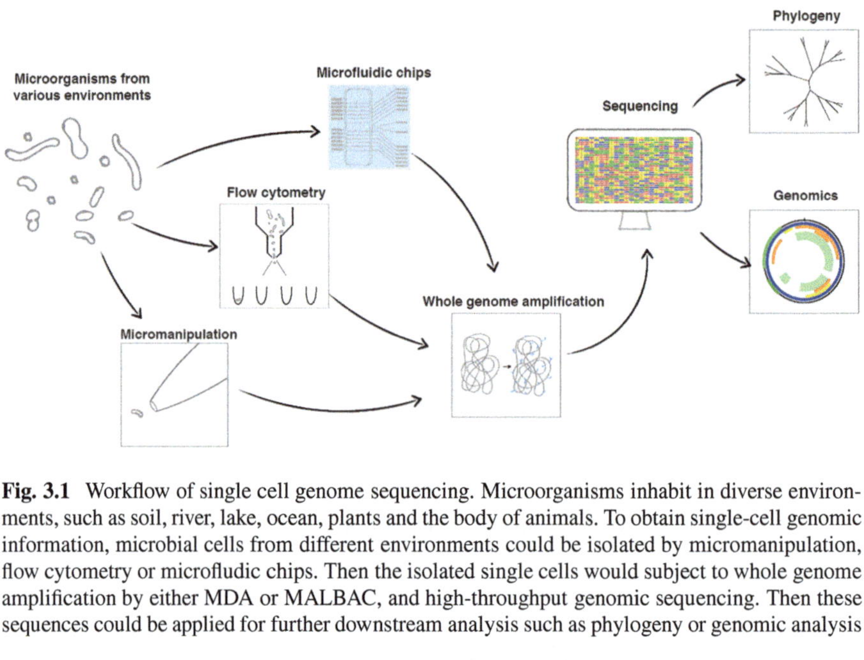

Fig. 3.1 Workflow of single cell genome sequencing. Microorganisms inhabit in diverse environments, such as soil, river, lake, ocean, plants and the body of animals. To obtain single-cell genomic information, microbial cells from different environments could be isolated by micromanipulation, flow cytometry or microfludic chips. Then the isolated single cells would subject to whole genome amplification by either MDA or MALBAC, and high-throughput genomic sequencing. Then these sequences could be applied for further downstream analysis such as phylogeny or genomic analysis

3.2 Methods of Single Cell Isolation

There are three principal single-cell isolation strategies: micromanipulation, flow cytometry and microfluidics chips.

3.2.1 *Micromanipulation*

Micromanipulation is a precise but laborious method to obtain a single cell. With the aid of micromanipulator devices, micrometer levels of precision in movement can be achieved by manually in operations such as holding, injecting and cutting cells. Glass micropipettes, optical tweezers and laser microdissection are the major tools used to manipulate single cells under the microscope.

3.2.1.1 Cell Isolation by Glass Micropipette

In this method, a target cell is captured with a disposable glass micropipette and then transferred to chips or tubes for subsequent analysis. The glass micropipette can be fabricated by commercial puller devices to the desired diameter. As an example of the successful application of this method to single-cell genome sequencing, Woyke et al. obtained a complete genome of an uncultured bacterium, *Candidatus Sulciamuelleri* DMIN (Woyke et al. 2010). A major drawback of this method is the long distance to be covered in transferring cells from a growth or storage medium into a tube or microtiter plate (MTP) for molecular analysis. The amount of time thus required for cell transfer limits the throughput of the method. In addition, once the micromanipulator is taken out of the field of view of the microscope it is no longer possible to visually control correct transfer of the single cell or bacterium into a tube or MTP. As a result, some cells fail to transfer correctly into the bottom of a tube or MTP well.

3.2.1.2 Cell Isolation by Optical Tweezers

Optical tweezers (OT) are capable of trapping and manipulating nanometer and micron-sized dielectric particles by exerting extremely small forces via a highly focused laser beam. In principle, a single selected cell is fixed with the laser beam and is separated from the mixed culture by moving the computer-controlled microscope stage and transferred into a predetermined separation chamber on the slide. Although this method has been successfully used in the isolation and culture of thermophilic bacteria and archaea (Huber et al. 1995), it has not been used for whole genome amplification and sequencing. In addition, this method may damage the target cells due to heating and photo damage, as power intensities rise to megawatts per square centimeter at the highly focused spot used for OT (Dholakia and Reece 2006).

3.2.1.3 Cell Isolation by Laser Microdissection

Laser microdissection and pressure catapulting (LMPC) is a method for isolating specific cells of interest from microscopic regions of tissue or cell samples or organisms. The method was firstly described by Schütze et al. (1998). Cells are spread on a polyethylene membrane, and localization of the target cells, based upon their morphological or histological criteria, is performed under microscopic visualization. Then, the surrounding membrane of the target cells are cut by laser dissection. After microdissection, a laser shot of increased energy is used to catapult target cells and the surrounding membrane into a common microfuge tube positioned above the sample for genomic amplification. Cells for LMPC sorting must be suspended in deionized water and dried on a polyethylene naphthalate membrane to prevent the formation of salt crystals that can cause problems for laser microdissection and cell localization this method can isolate whole specific cells or even a single chromosomal region by cutting away the unwanted parts (Thalhammer et al. 2004). In addition to genome amplification, available downstream applications include DNA genotyping and loss-of-heterozygosity analysis (Orba et al. 2003), RNA transcript profiling (Trogan et al. 2002), cDNA library generation, proteomics discovery and signal-pathway profiling. Although it offers many advantages in terms of speed, ease of use, and versatility, similar to micromanipulation, LMPC has the complication that proper cell placement into tubes or MTPs is difficult to control.

3.2.2 Flow Cytometry

In flow cytometry, cells are suspended in a stream of fluid and passed by an electronic detection apparatus, allowing both analysis and sorting of up to thousands of particles per second based on multiple physical and chemical characteristics. A flow cytometer provides "high-throughput" (for a large number of cells) automated quantification of set parameters of the cell. Among many examples of the successful application of flow cytometry is the discovery of many unexpected metabolic features of uncultivated microorganisms (Rinke et al. 2013).

Fluorescence-activated cell sorting (FACS) is a method that combining fluorescence cell staining with flow cytometry, which enable the single-cell sorting of biological cells from a heterogeneous mixture of biological cells. However, recent studies have shown that samples preserved in paraformaldehyde or on which fluorescence in situ hybridization has been performed are unsuitable for single-cell genomic amplification and sequencing (Clingenpeel et al. 2014).

Flow cytometry suffers several limitations. First, the cells must be in a single-cell suspension, posing a problem in the case of microbial cells that grow in form biofilms. The parameters per cell that can be measured simultaneously is limited by the number of detectors that can be used at the same time. In practice, this number is less than two dozen. The validation of results requires the simultaneous detection of multiple markers to increase specificity or, for example, the analysis of stimulated cytokine production of isolated cells to further confirm their phenotype. There is a surprising lack of standardization in assay and instrument set-up for flow cytometry. Standards are also lacking for how flow data are analyzed and reported. Lastly, because of the massive amount of data generated, flow cytometry data analysis can become very complicated and relies heavily on gating by a human expert.

3.2.3 Microfluidic Chips

Microfluidic chips provide a useful interface for the manipulation of single cells. Cell separation and sorting on a microfluidic chip can be achieved using a variety of microscale filters and fluid dynamics mechanisms, including field-flow fractionation, hydrodynamic filtration, and inertial microfluidics (Gossett et al. 2010). This method has been intensively applied on analysis of blood cells (Yu et al. 2014).

The major challenge in cell sorting by microfluidic chip is to design and fabricate chips for different samples. Complications may include heterogeneous populations of cells and the presence of noncellular particles, such as sediments and minerals.

3.3 Whole Genome Amplification Methods

Genome sequencing requires micrograms of DNA; however, single cells contain only picograms. Therefore, various methods of whole genome amplification (WGA) have been developed. Modified polymerase chain reaction (PCR) is the classic WGA method. This method requires thermocycling, random primers, degenerate or universal primers, and Taq DNA polymerase or similar enzymes. Taq DNA polymerase lacks 3′–5′proofreading activity and hence has high error rates. Newer WGA methods, multiple displacement amplification (MDA) and multiple annealing and looping-based amplification cycles (MALBAC) have provided improvements over PCR.

3.3.1 Multiple Displacement Amplification (MDA)

Multiple displacement amplification (MDA) is a non-PCR based DNA amplification technique. MDA still uses random primers; however, this method amplifies gDNA without thermocycling and generates larger products with a lower error frequency compared with conventional PCR amplification techniques (Lizardi et al. 1998). This method enables the rapid amplification of samples with very small amounts of DNA samples, providing a sufficient amount for genomic analysis.

The reaction starts by annealing random hexamer primers to the template: DNA synthesis is carried out at a constant temperature (30 °C) by a high fidelity enzyme, preferentially 29 DNA polymerase. This enzyme readily synthesizes DNA strands of 0.5 Mb length (Baner et al. 1998), and its high fidelity and 3′–5′ proofreading activity reduces the amplification error rate to 1 in 10^6–10^7 bases, compared to the reported error rate for conventional *Taq* polymerase of 1 in 9,000 (Tindall and Kunkel 1988). To achieve amplification, 29 DNA polymerase (Jiang et al. 2005) appears to initiate multiple replication forks on each circle and perform exponentially cascading strand displacement amplification.

MDA generates sufficient yield of DNA products for sequencing from a single cell and is therefore a powerful tool. The large size of MDA-amplified DNA products also provides desirable sample quality for identifying the size of polymorphic repeat alleles. Its high fidelity also makes it reliable enough to be used in single-nucleotide polymorphism (SNP) allele detection. Due to the strand displacement that occurs during amplification, the amplified DNA has sufficient coverage of the source DNA, providing a high quality product for genomic analysis (Spits et al. 2006). The products of displaced strands can also be subsequently cloned into vectors to construct libraries for sequencing (Zhang et al. 2006). These advantages make MDA the most widely used method for WGA.

The major drawback of MDA is amplification bias. Most studies on MDA have reported that this issue occurs due to over-amplification and allelic dropout. Another reported issue is that primer-primer interactions result in a sequenced product even in the absence of input template during MDA amplification. Therefore, there are problems regarding negative controls in the MDA reaction.

3.3.2 Multiple Annealing and Looping Based Amplification Cycles (MALBAC)

MALBAC is a PCR-based genome amplification method that introduces a step of quasilinear preamplification to reduce the bias associated with nonlinear amplification. In the preamplification phase, single-cell genomic DNA is melted at 94 °C and then annealed randomly with MALBAC primers at 0 °C, synthesizing semiamplicons (Lu et al. 2012). In the subsequent five temperature cycles, full amplicons are generated by a series quenching at 0 °C, extension at 65 °C, melting at 94 °C and self-looping at 58 °C with DNA polymerase. Self-looping of the full amplicons at the end of every cycle prevent these full amplicons from being used as a template for amplification during MALBAC, thereby reducing the amplification bias that is commonly associated with the uneven exponential amplification of DNA fragments by PCR (Zong et al. 2012). After the preamplification, only the full amplicons can be exponentially amplified in the following PCR using the common 27-nucleotide sequence as the primer. The PCR reaction will generate microgram levels of DNA material for sequencing experiments (Lu et al. 2012).

MALBAC has resulted in many significant advances over MDA amplification. MDA does not utilize DNA looping and amplifies DNA in an exponential fashion, resulting in bias. Amplification bias results in low coverage of the genome (Navin et al. 2011). The reduced bias associated with MALBAC has provided better genome sequence coverage, lower incidence of false positive and lower false negative mutations than other single-cell sequencing methods. However, the DNA polymerase used in the first cycle is error prone and can introduce sequencing errors that are propagated to the product DNA (Zong et al. 2012).

3.4 The Application of Single-Cell Sequencing on Uncultured Microorganisms

Archaea and bacteria play significant roles in global biogeochemical cycles (Falkowski et al. 2008), have important influences on human health (Turnbaugh et al. 2007), and contain a wealth of untapped novel natural products (Haefner 2003). Although there is a large amount of knowledge about microorganisms from years of study by traditional methods, the phylogeny and function of most microorganisms remain unknown. Single-cell sequencing has provided a novel way to analyze microbial community composition, diversity and potential metabolic capacity from the genome level.

During the last decade, an extensive portion of uncultivated microbial genomes that were previously inaccessible have been identified in the soil, ocean, deep subsurface, human body and other environments. Many of these achievements have made use of flow cytometry because of its high-throughput advantage over other single-cell isolation methods.

For example, single-cell genomic analysis of 201 uncultured archaeal and bacterial cells uncovered the presence of archaea-like metabolic functions in bacteria, and special features of bacteria in archaea, challenging the established boundaries between archaea and bacteria (Rinke et al. 2013). Using single consortia separation and sequencing, it was demonstrated that an uncultivated anaerobic methane oxidizer ANME-2a likely performs anaerobic oxidation of methane (AOM) through a complete reversal of methanogenesis from CO_2 reduction without involvement of canonical hydrogenase. ANME-2a has been shown to possess versatile electron transfer pathways (Wang et al. 2014). Genomic comparisons of different ANME groups could lay the foundation for understanding the environmental niche differentiation, physiology and evolution of different ANME subgroups.

Lloyd et al. performed single-cell sequencing of the predominant sedimentary archaea, including the uncultured marine benthic group-D (MBGD) and a miscellaneous crenarchaeotal group (MCG) which was recently proposed as a novel archaeal phylum "*Candidatus* Bathyarchaeota" (Lloyd et al. 2013; Meng et al. 2014). It was demonstrated that the predominant archaea in the marine sediment may have a previously undiscovered role in anaerobic protein remineralization (Lloyd et al.

2013). Moreover, single-cell genomics also have helped to uncover hundreds of co-existing subpopulations in wild marine cyanobacteria *Prochlorococcus* (Kashtan et al. 2014).

Recently, single-cell genomes from other species in bacterial environments, human commensals and opportunistic pathogens. For example, the candidate phylum TM6, *Porphyromonas gingivalis*, were obtained from a biofilm in a hospital sink drain (McLean et al. 2013a, b). These near-complete genomes will serve as important resources for the investigation of low-abundance pathogens and their transmission between the environment and the host (McLean et al. 2013b).

3.5 Conclusions

Single-cell sequencing is a fast-growing new technology that has been proven to be a powerful approach to gain new genomes of uncultivated microorganisms. Many taxonomic groups have only recently had their reference genome published due to the rapid progress in single-cell sequencing. In addition to single-cell sequencing of microbial cells, the sequencing of single eukaryotic cells is also improving and has exciting prospects for research into human cell development and disease. Looking to the future, we anticipate an exciting era of single-cell biology (McLean et al. 2013b).

Acknowledgements The authors thank Natural Science Foundation of China (NSFC, grant No. 91228201, 31290232), National High Technology Research and Development Program of China (Grant No. 2012AA092103-2) for their financial support.

References

Baner J, Nilsson M, Mendel-Hartvig M, Landegren U. Signal amplification of padlock probes by rolling circle replication. Nucleic Acids Res. 1998;26(22):5073–8.

Clingenpeel S, Schwientek P, Hugenholtz P, Woyke T. Effects of sample treatments on genome recovery via single-cell genomics. ISME J. 2014;8(12):2546–9.

Dholakia K, Reece P. Optical micromanipulation takes hold. Nano Today. 2006;1(1):18–27.

Falkowski PG, Fenchel T, Delong EF. The microbial engines that drive Earth's biogeochemical cycles. Science. 2008;320(5879):1034–9.

Gossett DR, Weaver WM, Mach AJ, Hur SC, Tse HT, Lee W, et al. Label-free cell separation and sorting in microfluidic systems. Anal Bioanal Chem. 2010;397(8):3249–67.

Haefner B. Drugs from the deep: marine natural products as drug candidates. Drug Discov Today. 2003;8(12):536–44.

Huber R, Burggraf S, Mayer T, Barns SM, Rossnagel P, Stetter KO. Isolation of a hyperthermophilic archaeum predicted by in situ RNA analysis. Nature. 1995;376(6535):57–8.

Jiang Z, Zhang X, Deka R, Jin L. Genome amplification of single sperm using multiple displacement amplification. Nucleic Acids Res. 2005;33(10):e91.

Kashtan N, Roggensack SE, Rodrigue S, Thompson JW, Biller SJ, Coe A, et al. Single-cell genomics reveals hundreds of coexisting subpopulations in wild Prochlorococcus. Science. 2014;344(6182):416–20.

Lizardi PM, Huang X, Zhu Z, Bray-Ward P, Thomas DC, Ward DC. Mutation detection and single-molecule counting using isothermal rolling-circle amplification. Nat Genet. 1998;19(3): 225–32.

Lloyd KG, Schreiber L, Petersen DG, Kjeldsen KU, Lever MA, Steen AD, et al. Predominant archaea in marine sediments degrade detrital proteins. Nature. 2013;496(7444):215–18.

Lu S, Zong C, Fan W, Yang M, Li J, Chapman AR, et al. Probing meiotic recombination and aneuploidy of single sperm cells by whole-genome sequencing. Science. 2012;338(6114): 1627–30.

Marcy Y, Ouverney C, Bik EM, Losekann T, Ivanova N, Martin HG, et al. Dissecting biological "dark matter" with single-cell genetic analysis of rare and uncultivated TM7 microbes from the human mouth. Proc Natl Acad Sci U S A. 2007;104(29):11889–94.

McLean JS, Lombardo MJ, Ziegler MG, Novotny M, Yee-Greenbaum J, Badger JH, et al. Genome of the pathogen Porphyromonas gingivalis recovered from a biofilm in a hospital sink using a high-throughput single-cell genomics platform. Genome Res. 2013a;23(5):867–77.

McLean JS, Lombardo MJ, Badger JH, Edlund A, Novotny M, Yee-Greenbaum J, et al. Candidate phylum TM6 genome recovered from a hospital sink biofilm provides genomic insights into this uncultivated phylum. Proc Natl Acad Sci U S A. 2013b;110(26):E2390–9.

Meng J, Xu J, Qin D, He Y, Xiao X, Wang F. Genetic and functional properties of uncultivated MCG archaea assessed by metagenome and gene expression analyses. ISME J. 2014;8(3): 650–9.

Navin N, Kendall J, Troge J, Andrews P, Rodgers L, McIndoo J, et al. Tumour evolution inferred by single-cell sequencing. Nature. 2011;472(7341):90–4.

Orba Y, Tanaka S, Nishihara H, Kawamura N, Itoh T, Shimizu M, et al. Application of laser capture microdissection to cytologic specimens for the detection of immunoglobulin heavy chain gene rearrangement in patients with malignant lymphoma. Cancer. 2003;99(4):198–204.

Rappé MS, Giovannoni SJ. The uncultured microbial majority. Annu Rev Microbiol. 2003;57(1):369–94.

Rinke C, Schwientek P, Sczyrba A, Ivanova NN, Anderson IJ, Cheng J-F, et al. Insights into the phylogeny and coding potential of microbial dark matter. Nature. 2013;499(7459):431–7.

Schutze K, Posl H, Lahr G. Laser micromanipulation systems as universal tools in cellular and molecular biology and in medicine. Cell Mol Biol (Noisy-le-grand). 1998;44(5):735–46.

Spits C, Le Caignec C, De Rycke M, Van Haute L, Van Steirteghem A, Liebaers I, et al. Whole-genome multiple displacement amplification from single cells. Nat Protoc. 2006;1(4):1965–70.

Thalhammer S, Langer S, Speicher MR, Heckl WM, Geigl JB. Generation of chromosome painting probes from single chromosomes by laser microdissection and linker-adaptor PCR. Chromosome Res. 2004;12(4):337–43.

Tindall KR, Kunkel TA. Fidelity of DNA synthesis by the Thermus aquaticus DNA polymerase. Biochemistry. 1988;27(16):6008–13.

Trogan E, Choudhury RP, Dansky HM, Rong JX, Breslow JL, Fisher EA. Laser capture microdissection analysis of gene expression in macrophages from atherosclerotic lesions of apolipoprotein E-deficient mice. Proc Natl Acad Sci U S A. 2002;99(4):2234–9.

Turnbaugh PJ, Ley RE, Hamady M, Fraser-Liggett C, Knight R, Gordon JI. The human microbiome project: exploring the microbial part of ourselves in a changing world. Nature. 2007;449(7164):804–10.

Wang F-P, Zhang Y, Chen Y, He Y, Qi J, Hinrichs K-U, et al. Methanotrophic archaea possessing diverging methane-oxidizing and electron-transporting pathways. ISME J. 2014;8(5):1069–78.

Woyke T, Tighe D, Mavromatis K, Clum A, Copeland A, Schackwitz W, et al. One bacterial cell, one complete genome. PLoS ONE. 2010;5(4):e10314.

Yu ZT, Aw Yong KM, Fu J. Microfluidic blood cell sorting: now and beyond. Small (Weinheim an der Bergstrasse, Germany). 2014;10(9):1687–703.

Zhang K, Martiny AC, Reppas NB, Barry KW, Malek J, Chisholm SW, et al. Sequencing genomes from single cells by polymerase cloning. Nat Biotechnol. 2006;24(6):680–6.
Zong C, Lu S, Chapman AR, Xie XS. Genome-wide detection of single-nucleotide and copy-number variations of a single human cell. Science. 2012;338(6114):1622–6.

Ying Chen Dr. Chen got his Ph.D. degree in biology at Shanghai Jiao Tong University in 2014. During his postgraduate studies he uncovered the physiology and evolutionary significance of two nitrate respiration system possessed by *Shewanella* species. And then he participated in the study of anaerobic methanotrophicarchaea by using the methods of single cell sequencing, nano-scale secondary ion mass spectrometry, fluorescence in situ hybridization and other culture-independent methods. Currently he is working at Guilin Medical University as a lecturer and research scientist, with interests in the microorganisms within the Karst landscape.

Fengping Wang Dr. Wang is a professor at School of Life Sciences and Biotechnology, Shanghai Jiaotong University (SJTU). Her major work interests are focusing on Phylogeny and Functions of uncultivated microbial organisms in marine environments. She is particularly fascinated by the uncultivated archaea in the deep subsurface marine sediments and rocks (the so-called "deep biosphere"), their diversity, phylogeny, adaptation, evolution, and ecological roles. Microorganisms are known to serve important roles in global biogeochemical cycles, yet the phylogeny and functions of most of them remain unknown. Recently, Dr. Wang and her team members utilized a variety of cross-disciplinary methods including the single cell sequencing technology on understanding the ecological functions of important yet-uncultivated archaea. They discovered the metabolic and electron transferring pathway of an uncultivated anaerobic methane oxidizing archaea ANME2a, revealed clearly for the first time the bio-mineralization mechanism of this methane oxidizing archaea. Meanwhile, a novel archaeal phylum named as "*Candidatus* Bathyarchaeota" was proposed by Dr. Wang and her co-workers which will draw attention and research interests on these novel organisms.

Chapter 4
The Application of Single-Cell Sequencing in Dynamic Transcriptomes

Kenian Chen, Xiaomin Dong, and Jiaqian Wu

Abstract There has been long-standing interest in developing procedures for profiling single-cell gene expression dynamics. Single-cell transcriptome profiling is useful for studying heterogeneity and mosaicism in tissues and understanding cellular behaviors and compositions. Recent advancements in RNA sequencing technologies dramatically improved sensitivity and specificity of gene expression quantifications, leading to rapid progress in single-cell transcriptome profiling methods. In this chapter, we provide a timely review on basic principles of single-cell RNA sequencing methods and discuss the technical challenges and their applications in research and medicine.

Keywords Application • Single-cell • Sequencing • Dynamic • Transcriptome

4.1 Introduction

Interest in single-cell transcriptome analysis, particularly for profiling rare or heterogeneous populations of cells, has been growing rapidly in recent years. The progress in single-cell transcriptome sequencing has been accelerated, largely due to the advancements of sequencing technologies and their successful adoption in various applications. As these technologies become mature, their prevalent use for studying heterogeneity and mosaicism in healthy and pathologic tissues, rare cell samples and many other applications is foreseeable. In this chapter, we will discuss the motivations, the basic principles, technical challenges of single-cell transcriptome sequencing, and its applications that will likely transform basic biological research and future medicine.

Why we need single-cell analysis technology while we already have technologies that can profile gene expression at tissue/cell-population level? Each single cell is unique as it occupies a specific space in tissue, carries its own copy of genome and

K. Chen • X. Dong • J. Wu (✉)
The Vivian L. Smith Department of Neurosurgery, Center for Stem Cell and Regenerative Medicine, Bentsen Center for Stroke Research, The University of Texas Medical School at Houston, Houston, TX, USA
e-mail: jiaqian2009.wu@gmail.com

X. Wang (ed.), *Single Cell Sequencing and Systems Immunology*,
Translational Bioinformatics 5, DOI 10.1007/978-94-017-9753-5_4

transcriptome which is particularly programmed and subject to changes induced by external environmental signals. The ability to access gene expression level at single cell resolution is critical for a better understanding of cellular behaviors and compositions in developing, mature and pathological tissues. However, our understanding of transcriptomes has been limited to tissue/cell-population-level observations over the last several decades. People are used to thinking about expression differences across tissues/conditions in terms of fold changes. In this sense, the subsets of cells which experience dramatic changes are averaged out or buried by the present of far more nonresponsive cells. On the other hand, it is been evidenced that single cells often response to inductive cues in an all-or-none style, but when quantified across the cell population, these responses are observed as gradual changes (Novick and Weiner 1957). To address these problems, a long-standing interest has been to establish procedures that can be used for profiling transcriptome dynamics in single cells.

One aspect which people may be concerned about single-cell level analysis, including single-cell transcriptome profiling, is that the information at the 'system' (tissue/cell population) level will be lost. Indeed, single-cells are quite different from each other and functionally homogeneous. Each single-cell, by itself, can not represent the whole tissue/cell-population, thereby in this sense, the 'system' level information is lost. However, this can be easily circumvented by studying multiple cells within the same system. And by combining the results from many single cells, one can then build up the information at the system level. This is also one of the major goals of systems biology – break the system into elements and then sum the parts together to get the whole picture.

Many areas of biological research and medicine can be transformed by applying single-cell transcriptome technologies. Some fields, such as neuroscience, immunology, cancer and induced pluripotent cells (iPSCs), in which cells are highly heterogenous and bulk tissue/cell-population based approaches may be insufficient began to benefit from these new tools. Some other fields, for example, pre-implantation genetic diagnosis in which the sample is rare, precious and hard to culture, is also made feasible by using the new single-cell technology.

Although single-cell transcriptome sequencing is full of promise and has made significant progress in the recent years, it is still at an early stage of development. Going to the single-cell level is not trivial currently. There are many inherent challenges in the procedures of single-cell sequencing. Each step, from the isolation of cells, to the amplification of their genomes or transcriptomes, and finally to making sense of the data using bioinformatics tools, still needs lots of improvements and innovations. Another consideration is the cost. In order to build up information at the systems level, single cells typically need to be sampled at higher numbers than tissues do. Although the price for sequencing is decreasing dramatically, the cost is still high for a typical small laboratory. When making decision whether to carry out study at single-cell level, one has to consider the benefit/cost ratio. Asking questions like "Does single-cell sequencing is really needed to meet the purpose of our study?" and "Given the same amount of budget, can we get much more insights from the single-cell data than just using bulk cell-population?" If the answer is no, one may not want to do single-cell sequencing immediately. It's more difficult and expensive, and is hard to interpret the variability in the results that one will encounter with currently available bioinformatics tools. If one is working with tissues that are known to be homogenous, single-cell sequencing might not have significant advantages compared to studies with bulk tissue approaches. However, for more complex systems such as CNS (central nervous system), blood and immune system or even cell types that comprise them, single-cell sequencing may provide advantages if sufficient funding is available.

4.2 An Overview of Single-Cell Transcriptome Sequencing

While in the field more than one single-cell sequencing protocols exist, the fundamental principle and components behind them are similar. Before giving a detailed look at each part of the single-cell transcriptome sequencing process, we will review the fundamental parts of typical workflow in this section (Fig. 4.1).

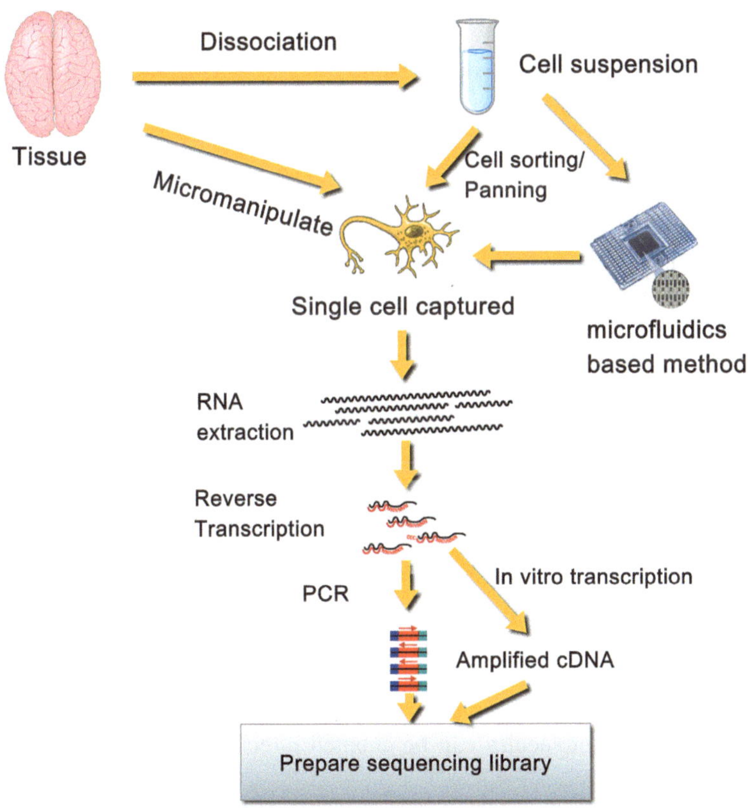

Fig. 4.1 Overview of single-cell RNA sequencing
The basic components of single-cell RNA sequencing are illustrated

4.2.1 Obtaining Single Cells from Tissue/Cell-Population

It is obvious that for single-cell RNA sequencing, we need to isolate single cells from a large cell population first, and put each of them into individual reaction systems. There exist many methods for separating single cells from large population, even before high-throughput expression profiling technologies were invented. The first one is micromanipulation using micromanipulator. A micromanipulator is a device which is used to physically interact with cell samples under a microscope. Coupled with micro-capillaries, micromanipulator can hold, and extract a cell's contents directly. Micromanipulation is precise but labor intensive and time consuming. It requires manual operation, so is also very low-through-put. For tissues that can be dissected to cell suspensions, cell sorting technologies (e.g. FACS/MACS) can help to handle and enrich cells expressing specific markers automatically. However, such methods do not fully isolate each cell from one another, thus are not compatible with downstream single-cell gene expression analysis, where the cellular contents from each cell must be kept isolated after cell lysis. The last one and most promising one is the recently developed microfluidic devices. This kind of devices utilizes liquid flows in small volumes and can obtain single cells in a high-throughput manner, and downstream expression profiling can also be integrated into such a device. The details of such a device will be discussed in Sect. 4.3.

4.2.2 RNA-Sequencing

After single cells are obtained, the RNAs will be extracted and used for sequencing library preparation. Many RNA sequencing protocols are available for single-cell expression profiling, either from academic or commercial sector. Although the methods vary from one another, they all begin with the conversion of RNA to the first strand of complementary DNA (cDNA) utilizing reverse transcriptase. These methods have a common goal to capture the original RNA molecules and amplify them as evenly and accurately as possible. In practice, capture efficiency is influenced by reverse transcriptase's ability of sampling RNA molecules from the cell. This sampling process is stochastic and can be improved by reducing reaction volumes and using more efficient reverse transcriptase. Amplification step is also critical, as any bias in the initial step can be magnified and make the resulting data noisier. Minimizing the number of PCR cycles, increasing the starting material by pooling multiplexed barcoded cDNA together are the considerations that can reduce amplification bias. Some methods also utilize unique molecular index sequences to label individual RNA molecules, similar to digital PCR, subsequently the absolute number of original molecules can be counted directly to avoid the influence of uneven amplification.

Scaling up is also important, as many single-cells should be sampled in one experiment to ensure the 'system-level' information such as biological variability can be captured. Microfluidics devices can be designed and integrated with downstream sequence library preparing process, thus provide high throughput and standardized handling.

4.2.3 Bioinformatics Analysis

As the single-cell RNA sequencing technology keeps developing, more data will be produced in the near future. This ‘data explosion’ challenge also needs to be addressed with the help of newly developed algorithms and software. Wet lab biologists will need to collaborate closely with computational biologists to interpret the large data sets generated from numerous single-cell experiments. Statistical methods to distinguish true biological variability from technical noise in single-cell transcriptome sequencing are also under development.

Currently, a standard pipeline for data analysis is not established yet, as data come from different platforms with different experimental designs. However, the low-level data processing is essentially the same as standard RNA-Seq and some mature algorithms/software/packages can be applied directly. This low-level data processing includes read mapping and transcriptome construction.

4.3 The Development of Microfluidic Devices for Single Cell Sequencing

As we have mentioned in the previous section, the initiative step of single cell sequencing is to separate single-cells from a large cell population. In this section, we will discuss some methods for single-cell analysis (not only for sequencing). Then we will focus on microfluidic devices as they hold future promise.

Methods for single-cell analysis have been evolving for a long time. Microscopy is undoubtedly the first device for single-cell analysis, since cells were first discovered under a microscope. Along the way, other methods include micromanipulator, histological staining, in situ hybridization (ISH), patch-clamp, mass cytometry, and flow cytometry were developed and people turn to these methods gradually (Batish et al. 2011; Bendall et al. 2011; Gaynor et al. 1996; Krutzik et al. 2008; Krutzik and Nolan 2006; Spruce et al. 1989; Evans and Yeung 1989).

All these methods have their own strength and weakness. Micropipette is extremely low throughput and labor intensive. Histological staining and in situ hybridization are sensitive and can be used to detect multiple genes simultaneously with the special information conserved, but still require intensive labor for preparation procedures.

One of the common ways for obtaining single cells is to use a Flow Cytometry device. This device can separate cell population into single cells in a number of ways, depending on how the cells will be used later. The first way is sorting the cells based on their size (Ivanov et al. 2008). In this way, only the smallest or largest cells can be sorted out. The second way is separating cells based on their shape and morphology (Kastrinaki et al. 2008). The third way is using Fluorescence-activated cell sorting (Joo and Nam 2009) or briefly FACS, which can separate the cells based on the fluorescent labels tagged on a given cell type. FACS is high throughput in cell number and its multiplexing ability (the ability to obtain multiple cell types simultaneously) has increased in recent years (Perfetto et al. 2004). Cells are usually tagged with fluorescent antibodies, which can bind to either trans-membrane or intracellular proteins, or with fluorescently labeled probes targeting specific mRNA or DNA molecules. Multiple markers can be used in combination and be simultaneously detected. The adoption of FACS has allowed for the identification of cell subpopulations and is important in enrichment of samples or cell collection.

Although flow cytometry has many merits, it is not flawless. Sample preparation is still labor intensive and time consuming. The multiplexing ability of flow cytometry is limited by fluorescence spectral overlap. Tagging of targets is performed directly on whole cells and some molecules within the cell may be difficult to probe. Some secreted proteins are also not possible to be analyzed using flow cytometry.

In recent years, the development of microfluidic devices emerged for the purpose of single-cell analysis (Bontoux et al. 2008; Fan et al. 2008, 2011; White et al. 2011; Di Carlo and Lee 2006; Chao and Ros 2008; Köster et al. 2008; Marcus et al. 2006; McKenna et al. 2011; Wheeler et al. 2003). Microfluidic devices utilize liquid flows in small volumes and offer many advantages amiable for single-cell analysis. Such devices, usually contains compartments with dimensions in the same range as mammalian cells, allow for better controlled and precise manipulation of single cells while maintaining reagent volumes minimum. The miniature environment microfluidic device provided allows the reduction of reagent volume, increased concentration, and avoids loss of samples and contaminations. In addition, microfluidic devices can be fabricated to incorporate multiple functional compartments, enabling cell capture, lysis and downstream analysis within the same device (Manz et al. 1990). Briefly, microfluidics devices can be characterized as miniature assembly lines in which reactions take place as the samples circulate from one specific functional compartment to another through capillary channels. The devices can be embedded in carved silicon platforms similar to microarray chips. These chips are designed with the help of computer and fabricated using systems similar to the production of Micro Electro Mechanical Systems (MEMS) (Kauzlarick 2003). Chips can be classified into two types according to the strategy used for creating isolated compartments. One type uses solid borders materials such as microchambers, microwells or microvalves to separate functional compartments; the other type uses fluid borders, e.g. microemulsions (Schmid et al. 2010). For more information about microfluidic devices, readers can refer to (Nguyen and Wereley 2006; Bruus 2008; Kaajakari 2009).

4.4 RNA Sequencing in Details

RNA sequencing becomes an established method in recent years and is rapidly replacing microarray as the method of choice for transcriptome profiling (Sultan et al. 2008). In addition to superior accuracy in the expression level quantification, RNA sequencing offers other advantages, such as novel transcripts detection, splicing isoform construction and strand-specific expression profiling (Ozsolak and Milos 2011). Ideally, RNA sequencing should cover the entire transcript evenly, retains strand information and allow accurate quantification of mRNA expression levels. However, depending on the experimental protocol used, particularly for single-cell RNA sequencing, these expectations are not simultaneously possible with currently available methods (Hebenstreit 2012).

Most work of a sequencing project is about the sequencing library preparation, before putting samples on a sequencing machine. Sample processing for preparing RNA sequencing library is largely based on molecular biological protocols that were established decades ago. Most RNA sequencing library preparation methods include the following basic steps: poly(A) + RNA isolation, fragmentation and size selection, reverse transcription and amplification.

4.4.1 Poly(A) + RNA Selection

The purpose of poly(A) + RNA selection step is to remove structural RNAs such as ribosomal RNA and tRNA from sequencing library. These structural RNAs represent the majority of cellular RNA and can consume a large number of sequencing reads, thus reduce the information gained. Most mRNAs contain a poly(A) tail, but structural RNAs do not (Kingston 2001). Thus Poly(A) selection enriches mRNA, and is useful when analyze low-abundance mRNAs. Briefly, total RNA is denatured to expose poly(A) tails. Poly(A) + RNA is then bound to oligo(dT) which is covalently attached to a given substrate. Subsequently poly(A)- RNA is washed away and poly(A) + RNA is eluted for later processing. Although its advantages is obvious, this method precludes poly(A)- RNA (Marzluff et al. 2008). Advancements in sequencing depth might finally make poly(A) selection unnecessary (Tariq et al. 2011).

4.4.2 Fragmentation and Size Selection

Currently, most next generation sequencing machineries can only sequence short reads. The fragmentation step is thus necessary to produce short RNA or DNA fragments from the original transcripts, in order to achieve a suitable fragment length for the sequencing machine. An optimized fragmentation procedure also allows better quantification accuracy and reduces positional biases across the transcripts represented. Various methods have been successfully applied for fragmentation step, including mechanical, chemical and enzymatical methods. Since most sequencing machines have constrains on the DNA length, a gel or beads based size selection should also be performed.

4.4.3 Reverse Transcription

Reverse transcription is used to generate cDNA template for the amplification step. It is necessary because it is not clear whether RNA-dependent RNA polymerase is suitable for RNA amplification (Ortín and Parra 2006). Reverse transcriptase can begin polymerizing from inside (primed by random hexmers) or at the end (primed by olig(dT)) of the RNA template, depending on the type of primers used. Ideally, the randomness in many preparation steps should lead to fragments whose starting positions (relative to the transcripts which they come from) appear uniformly at random. However, it is not the case in practice. Both positional (local effect in which fragments preferentially locate towards either the beginning or end of transcripts) and sequence-specific (a global effect where the sequence surrounding both ends of potential fragments affects their likelihood of being selected) biases can be introduced during reverse transcription step. These biases are largely affected by the enzyme's processivity (Fairfield et al. 1983; Islam et al. 2011). Bias introduced when priming by random primers can be worsen as the enzyme drop-off decreasing. In contrast, bias introduced with fixed primer (e.g. $3'$ end of transcript) can be worsen as the enzyme drop-off increasing.

Biases can be reduced by using better enzymes and optimized reaction systems. Fragmentation step performed on the original RNA template is also considered as a method that can largely reduce biases; however, it may also lead to RNA molecules loss as RNAs are likely damaged during the processing. The ultimate solution might be amplification-free RNA sequencing which does not require reverse transcription and thus removes all these biases. Statics models in the later data processing stage can also be used to improve expression estimation.

4.4.4 Amplification

Most sequencing technologies require a large amount of starting material (nanogram quantities of DNA) which is usually not satisfied with a single-cell's cellular content (approximately 10 pg of total RNA with 0.1 pg of poly(A) + RNA). Substantial amplification is thus necessary. The challenge of amplification is to maintain the original relative abundances of RNA molecules through the whole process. The strand information about the transcripts will also be lost after the second strand synthesis, though strand specific protocols do exist to circumstance the situation.

Polymerase chain reaction (PCR) and in vitro transcription (IVT) are two available methods for amplification. Although PCR have been used in a number of single-cell studies for preparing sequencing libraries, it is worth knowing that biases in PCR efficiency for particular sequences can also be amplified exponentially. Most researchers believe that limiting the number of PCR cycles can reduce the bias. However, since the bias is due to the enzyme preference, it is likely sequence specific and variable with the gene expression. The degree of bias is not easily predictable.

In vitro transcription (IVT) of cDNA into RNA can be used to amplify RNAs. Since the number of templates are not changed during the process, a linearly amplification can be achieved in theory. However, it is not without problems. Amplification efficiency can exhibit sequence specificity and some sequences are dropped out. Amplified sequences are also shorter compared to PCR.

The development of amplification-free RNA sequencing might finally resolve all the problems that come with amplification. One such a method called FRT-seq has been reported by Lira et al. in 2010 but not tested in single-cell RNA sequencing yet (Mamanova et al. 2010; Mamanova and Turner 2011). The critical step which makes FRT-seq unique is that the reverse transcription reaction takes place on the Illumina flowcell using reverse transcriptase.

Another strategy to overcome amplification bias is to label first strand cDNA product with unique sequence tags (Kivioja et al. 2012; Shiroguchi et al. 2012; Fu et al. 2011; Casbon et al. 2011). After amplification, these unique tags can be used to count the number of original RNA molecules in the cells without being affected by the distortions from amplification. For tagging each cDNA copy of an individual RNA molecule uniquely and randomly, the diversity of tags should be large enough (in excess of RNAs that need to be tagged), and these tags should be able to tolerate some extent of sequencing errors in order to make sure they are not misidentified.

4.4.5 Currently Published Single-Cell RNA Sequencing Protocols

In this section, we will summarize several single-cell RNA sequencing methods that have been published in recent years.

In 2009, Tang et al. developed the first protocol for single-cell RNA sequencing and applied it to sequence a single mouse blastomere (Tang et al. 2009). They used micropipette to separate individual cells. After the cell lysis, reverse transcription were directly performed using poly(T) primer which contains an anchor sequence on the 3′ end. Free primers were removed with exonuclease. Poly(A) tails is then attached to the 3′ ends of the first strand by terminal transferase. Second strand synthesis was initiated using poly(T) primer with a different anchor sequence. Subsequently, PCR amplification was performed on the cDNAs using anchor sequence primers. Resulting cDNAs were sheared and size selected and then subjected to sequencing library preparation. Notably, since the start position of second strand synthesis is where the first strand synthesis stops, bias can be introduced depending on the drop-off of the reverse transcriptase (Fig. 4.2).

In 2012, Ramsköld et al. developed a protocol called SMART-Seq which improves both the average transcript size and the number of full-length transcripts

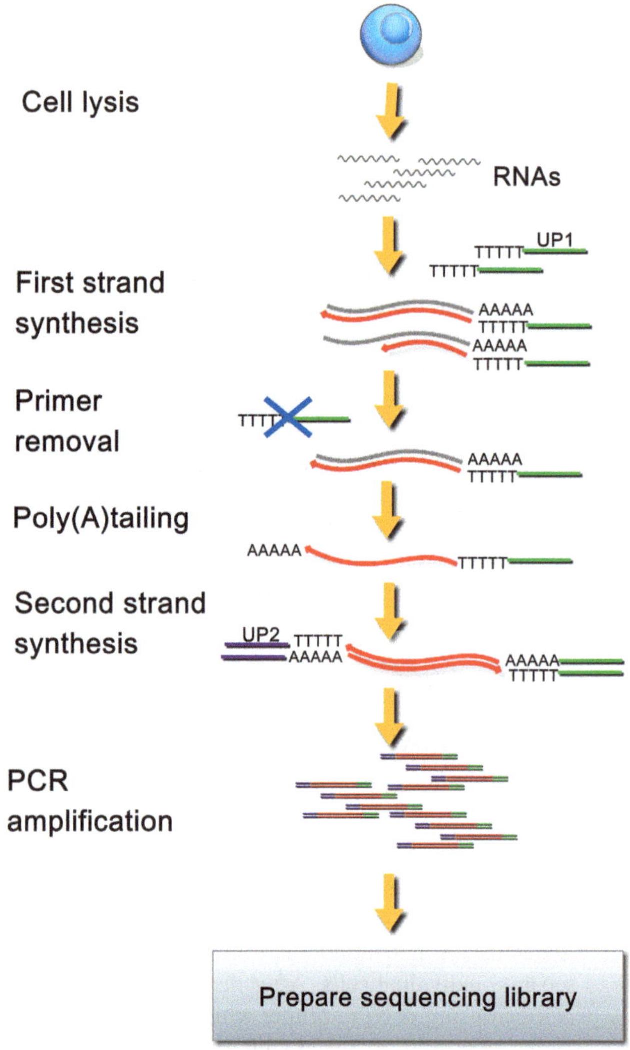

Fig. 4.2 Schematic of Tang's single-cell RNA sequencing method
After cell lysis, mRNAs are reverse-transcribed to cDNA using poly(T) primer which contains anchor sequence (UP1). Free primers are digested. Poly(A) tails are added to the 3′ end of first strand cDNA. Second strand cDNAs are synthesized using poly(T) primer with anchor sequence (UP2). After PCR amplification, cDNAs are subjected to sequencing library preparation

(Ramsköld et al. 2012). The major feature of this method is the usage of a reverse transcriptase from the Moloney murine leukemia virus (MMLV). This enzyme possesses two features: template switching and terminal transferase activity, which are critical to SMART-Seq. The enzyme can activate its terminal transferase ability and adds non-templated cytosine residues to the cDNA when it reaches the 5′ end of an mRNA (Zhu et al. 2001). If oligonucleotides with guanine residues which can base pair with the cytosines are present in the reaction system, the reverse transcriptase can then switch templates and start transcribe in reverse direction. This mechanism was called switch mechanism at the 5′ end of RNA templates (SMART), and so comes the name of SMART-Seq. Because template switching has a preference for 5′-capped RNA, the method enriches for transcripts that have an intact 5′ end, thus improves the coverage and number of full length transcripts. In 2013, Simone et al. published an improved version of SMART-Seq called Smart-seq2 (Picelli et al. 2013). This method, showed by the authors, has improved detection, coverage, bias and accuracy compared to SMART-Seq. One of the limitations of the SMART-Seq or SMART-Seq2 as the authors indicated is that, when starting RNA amount is low (10 pg for a single cell), some transcripts, particularly those expressed in lower level, exhibit random loss. That said, for low-abundance transcripts, the observed variability between cells was mainly technical variation, but for medium- and high-abundance transcripts, was mainly biological variation. For more than 10 cells, technical losses were small and robust. The cost-effective generation of single-cell RNA sequencing library using off-the-shelf reagents might facilitate the adoption of SMART-Seq2 in near future (Fig. 4.3).

Another method rely on template switching is called "Single-cell tagged reverse transcription" STRT which is published prior to SMART-Seq by Islam et al. (2012). The major difference is that the template-switching oligonucleotides contain a barcode sequence in addition to guanine residues. The barcoding allows the samples being pooled together after first-strand synthesis and sequenced later in a multiplex way (Fig. 4.4).

In 2012, Tamar et al. developed a protocol called CEL-seq (Hashimshony et al. 2012). The method is the first one that replaces PCR amplification with IVT. The oligo(T) primer for first strand synthesis was combined with barcode sequence and T7 promoter. After the first strand synthesis, the cDNA templates were in vitro transcribed into RNA using T7 polymerase which has the ability to repeatedly bind to the T7 promoter and transcribe. Because the number of template does not increase after each round of transcription, the amplification grows linearly. The authors demonstrated that CEL-seq has higher sensitivity and accuracy than STRT (Fig. 4.5).

In 2013, Yohei et al. developed yet another protocol named Quartz-seq (Sasagawa et al. 2013). A primer contains oligo-dT, T7 promoter and a PCR target region (M) sequence was used for the first cDNA strand synthesis. RT primer was then digested by exonuclease I. A poly(A) tail was subsequently added to the 3′ end of first-strand cDNA. Second-strand cDNA was synthesized using oligo-dT primer which contains a PCR target region. Later, suppression PCR was

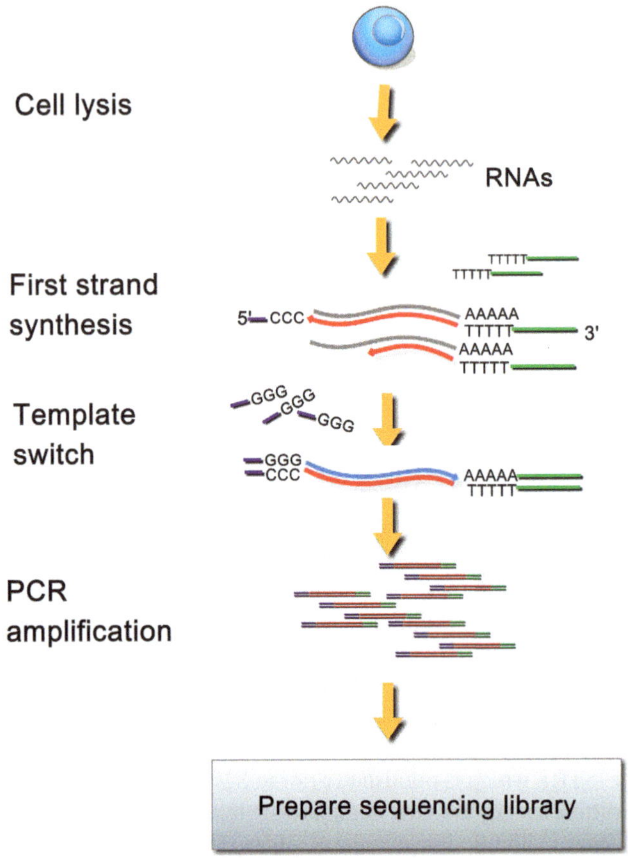

Fig. 4.3 Diagram of SMART-Seq method
After cell lysis, mRNAs are reverse-transcribed to cDNA using poly(T) primer by MMLV reverse transcriptase. When MMLV come to the 3′ end of first-strand cDNA non-templated cytosines are added. Oligonucleotides with guanines are used as primer and reverse transcription switches from RNA to DNA template and synthesis the second-strand cDNA. PCR amplification is then performed and cDNAs are subjected to sequencing library preparation

performed and products were purified to yield high-quality cDNAs for sequencing. The authors showed that Quartz-seq outperformed other methods in reproducibility and sensitivity (Fig. 4.6).

In early 2014, Saiful et al. published a single-cell sequencing method which utilized unique molecular identifiers (UMI: short sequence tags) to label individual molecules during reverse transcription (Islam et al. 2014). The numbers of original cDNA molecules were counted by enumerating the total number of distinct UMIs mapped to each genome position. The authors showed that utilizing of UMI can nearly eliminate amplification noise, and sample preparation using microfluidic and optimized reagents can improve mRNA capture efficiency fivefolds (Fig. 4.7, Table 4.1).

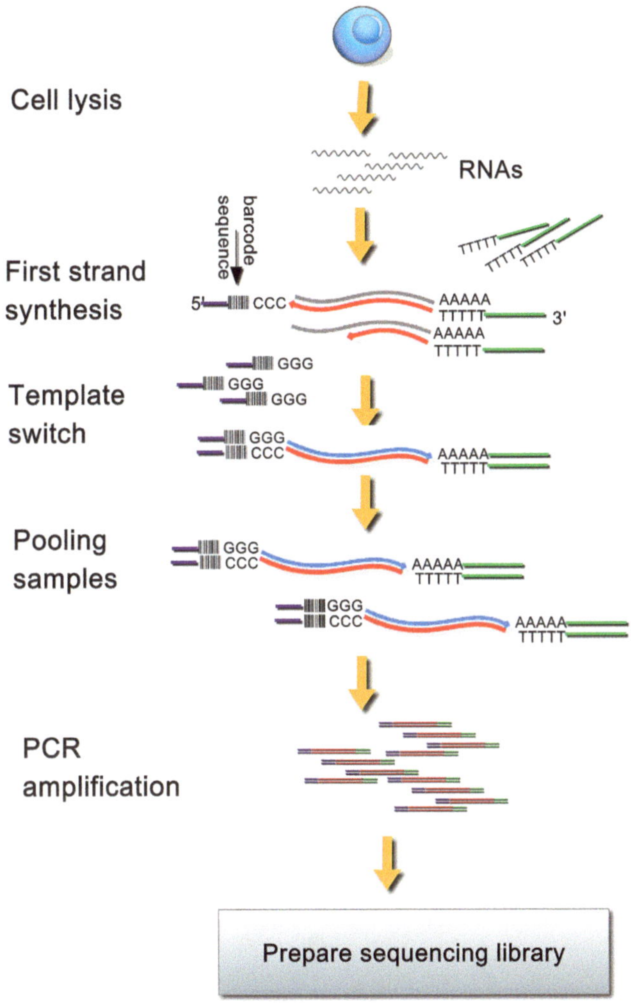

Fig. 4.4 Diagram of STRT method
STRT method is similar to SMART-Seq in utilizing the template-switch mechanism. STRT features a barcode sequence in the introduced template-switching oligonucleotides. After cDNA synthesis, samples can be pooled together and subjected to PCR amplification before sequencing library preparation

4.5 Bioinformatics Challenges in Analyzing Single-Cell Transcriptome Data

At the most fundamental level of analysis, a single-cell RNA sequencing experiment tells us the abundance of all transcripts within a cell represented as sequenced 'reads' that come from fragments of transcripts. To obtain the quantitative infor-

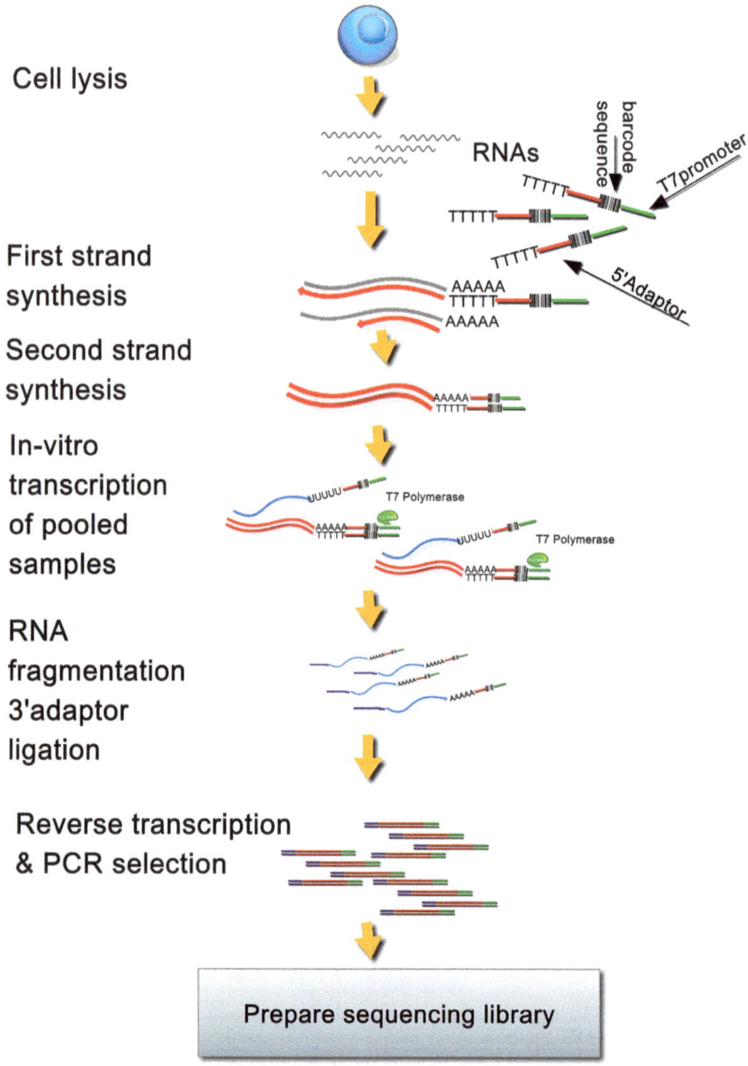

Fig. 4.5 Diagram of CEL-seq method
After cell lysis, primer with T7 promoter, oligo(T) and barcode sequence is used for reverse transcription. After second-strand synthesis, samples can be pooled together and in vitro transcription is performed using T7 polymerase. Amplified RNAs are then fragmented and purified for sequencing library preparation

mation out of these reads, mapping and transcriptome reconstruction must be performed first. Many algorithms and software packages are available for standard RNA sequencing experiment; with a little modification they can readily be adapted

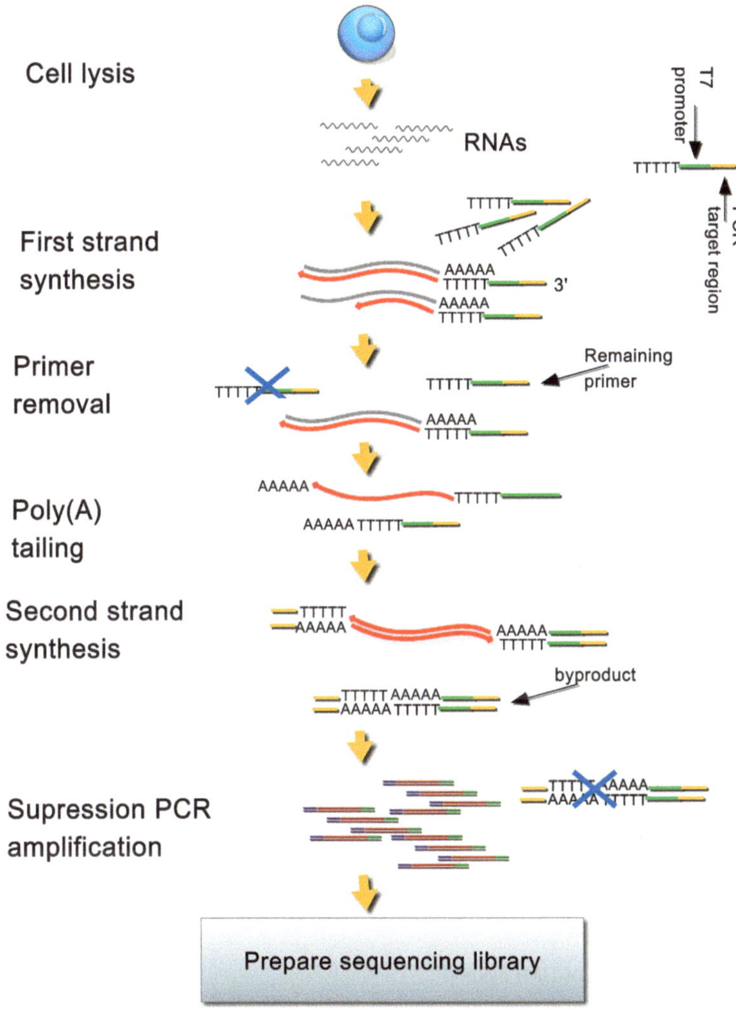

Fig. 4.6 Diagram of Quartz-seq method
First-strand cDNAs are synthesized using primer which contains oligo(T), T7 promoter and PCR target region sequences. Majority of RT primers are digested after first-strand synthesis, some primers survive though. Poly(A) tail is added to 3′ end of first-strand cDNA and surviving RT primer. After second-strand synthesis, byproducts from the surviving primers also formed. A suppression PCR is performed and byproducts amplification is suppressed. The resulting high-quality cDNA without byproducts is obtained and subjected to later sequencing library preparation

to single-cell data. An established workflow using the Tuxedo software package is shown in figure (Trapnell et al. 2009, 2010, 2012; Langmead et al. 2009) (Fig. 4.8).

Although the workflow for low-level data processing are similar between standard and single-cell RNA sequencing, the statistical assumptions for testing

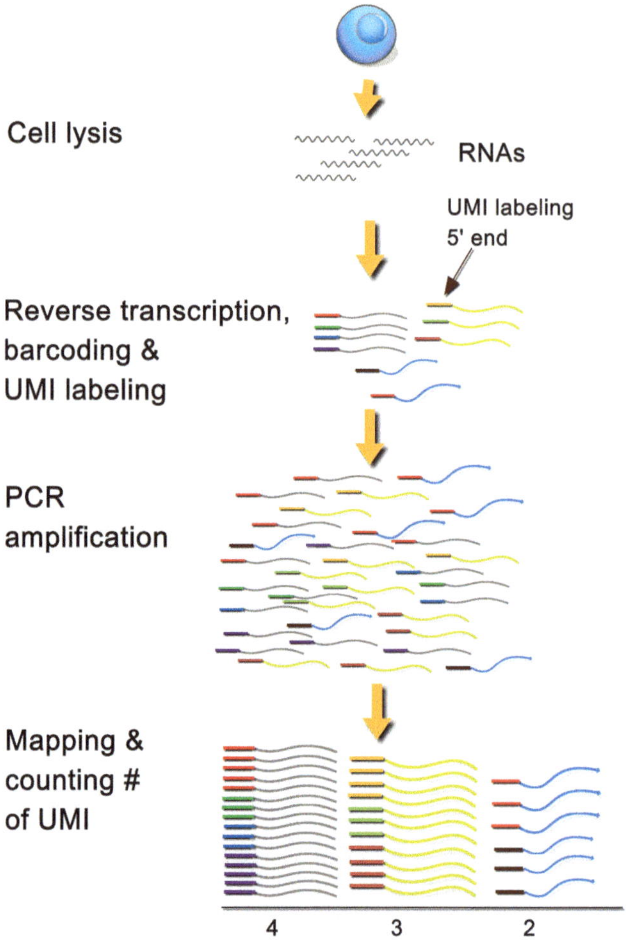

Fig. 4.7 Basic principle of UMI method
The UMI (unique molecular identifier) method labels individual cDNA molecules with short sequence as molecular identifiers. UMI go through the PCR amplification process with original cDNA. After sequencing, reads are mapped to the reference genome, the number of unique molecular identifiers that mapped to the same position correspond to the number of original RNA molecular come from that position

differential expression in standard RNA sequencing are not necessarily applicable (Islam et al. 2011; Dobson et al. 2004; Lovén et al. 2012). One of the challenges is the sequencing depth variation among single-cell samples, which contributes to the variation of sensitivity, e.g. shallow sequencing depth could loss detection of some transcripts. Another issue is the process and underlying assumption for normalization. In RNA sequencing data analysis, normalization is necessary to correct technical biases. Normalization procedures developed for standard RNA sequencing include rescaling by estimated sequencing depth in order to remove effects raised

Table 4.1 Summary of published single-cell RNA sequencing methods

	Tang's	SMART-seq	STRT	CEL-Seq	Quartz-Seq	UMI
First-strand synthesis	Yes	Yes	Yes	Yes	Yes	Yes
Remove fist-strand primer	Yes	No	No	No	Yes	No
Template switch	No	Yes	Yes	No	No	Yes
Amplification method	PCR	PCR	PCR	IVT	Suppression PCR	PCR
Molecular counting	No	No	No	No	No	Yes
Multiplexing	No	No	Yes	Yes	No	Yes
Reference	Tang et al. (2009)	Ramsköld et al. (2012)	Islam et al. (2012)	Hashimshony et al. (2012)	Sasagawa et al. (2013)	Islam et al. (2014)

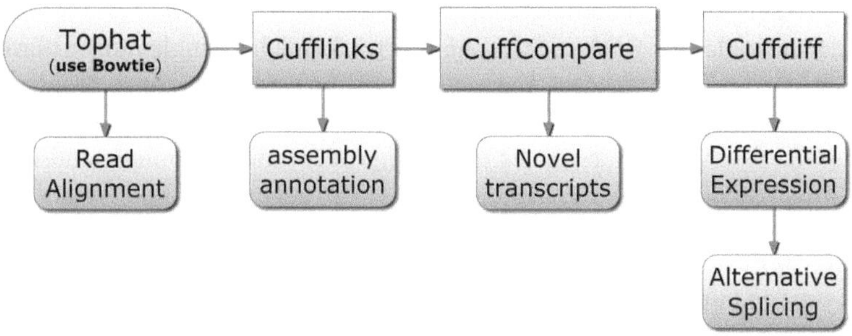

Fig. 4.8 Workflow for basic RNA sequencing analysis
An established workflow of standard RNA sequencing data analysis utilizes Tuxedo software package. Tophat (which use Bowtie as read mapper) is used for read mapping. Cufflinks is used for transcript assembly. Cuffcompare is used for novel transcription and cuffdiff for differential expressed gene calling

by different sequencing depth for individual samples (Anders and Huber 2010; Bullard et al. 2010; Li and Tibshirani 2011; Robinson and Oshlack 2010). The assumption underlying this normalization procedure is that the majority of genes are not differentially expressed between two samples compared, and different cell types has similar amount of RNA. This assumption is usually considered satisfied as for standard RNA sequencing, and has been accepted in decades of microarray experiments. However, for single-cell RNA sequencing, it would not be appropriate as we know that come down to the single cell level, the total transcript number varies significantly between individual cells.

Sequencing samples together with RNA spike-ins, for example those developed by External RNA Controls Consortium (ERCC), is useful for obtaining information about the relative efficiency, detection limits and technical noise of different amplification methods. Spike-in can also be used for normalization, so the differences of total transcript counts can be detected (Islam et al. 2014; Lovén et al. 2012; Wu et al. 2013; Jiang et al. 2011; Treutlein et al. 2014). Sequencing methods that utilize single molecule counting (such as UMI method) can calculate the original RNA molecule number directly and indicate the amplification level of individual RNA molecules. Obtain the original RNA molecule number per cell will also make the statistic test more straight forward.

4.6 The Applications of Single-Cell RNA Sequencing

Despite single-cell RNA sequencing technologies are still at an early stage in uncovering cell-to-cell variation in transcriptome regulation, recent studies which adopted these new tools have already given us new insights into development, cancer, immunology and neuroscience. We illustrate some applications of single-cell RNA sequencing that emerged in recent years.

The first single-cell RNA sequencing was performed on single mouse blastomere by Tang et al., demonstrating the power of whole-transcriptome profiling from single-cell using sequencing technology (Tang et al. 2009). In this study, the authors detected 75 % (5,270) more genes than using microarray technique and identified 1,753 novel splice junctions. Moreover, 8–19 % genes with known isoforms are found exhibiting alternative splicing in the same blastomere or oocyte, revealing the complexity of transcriptome in individual cells. In 2013, Yan et al. extended the method to 124 individual cells from human preimplantation embryos and embryonic stem cells (Yan et al. 2013). The authors discovered 2,733 novel lncRNAs (long non coding RNA), many of which are expressed at specific developmental stages. Within single cells, averagely, lncRNA can be expressed as high as ~40.5 % of the protein-coding gene, implying lncRNAs are relatively abundant within individual cells and are potentially important regulators.

In neuroscience filed, single-cell gene expression analysis is particularly useful, because the nervous system consists of the most diverse types of cells, especially neurons (Bota et al. 2003; Luo et al. 2008). Neurons can be categorized into different types based on their expressed molecule, morphology, electrophysiological characteristics, and connections (Esumi et al. 2008). Single-cell transcriptome profiling can provide a more panoramic view and has the potential to refine cell type definitions. Qiu et al. performed single-cell RNA sequencing on single neurons following electrophysiology recording (Qiu et al. 2012). The authors found that the correlation coefficient of gene expression between individual neurons in situ was much lower than the cultured neuronal cells. Thus, this study shows that morphologically identical neurons, even from the same brain region, can exhibit distinct gene expression patterns.

In 2013, Shalek et al. investigated heterogeneity in the response of mouse bone-marrow-derived dendritic cells (BMDCs) to lipopolysaccharide (LPS) using single-cell RNA sequencing (Shalek et al. 2013). They found extensive bimodal variation in mRNA abundance and splicing patterns that are not observed in previously studies using cell populations. Hundreds of key immune genes are bimodally expressed across cells (e.g. mRNA abundance has bimodal variation among individual cells), even for genes that are highly expressed in cell population. They also identified a module of 137 highly variable but still co-regulated antiviral response genes. Their study demonstrated the power of single-cell transcriptome in the discovery of functional diversity in cells and deciphering cell states and circuits.

Single-cell RNA sequencing has also been applied to cancer research. Cann et al. demonstrated such an application (Cann et al. 2012). Circulating tumor cells (CTC) mediate metastatic spread of many solid tumors. To derive additional information about tumors, the authors performed single-cell RNA sequencing on CTCs isolated from the blood of patients. Although the major purpose of the study is to access the ability of cell isolation methods, the authors also showed that transcriptional signatures of prostate tissue and of cancer can be readily detectable using single CTC RNA sequencing, indicating the potential power of single-cell sequencing in cancer diagnosis.

4.7 Conclusion

Individual cells are the fundamental units of almost all life forms. Therefore, the ability of single-cell analysis is not merely a step towards more sensitive measurements, but also a radical revolution of the way we used to understand biological phenomenon fundamentally. In this chapter, we discussed recent advances and challenges in single-cell RNA sequencing technologies. In addition, we illustrated how exploratory studies using these new technologies have led to new insights. Whole-transcriptome profiling at single cell level will resolve many longstanding debates in biology. If enough individual cells are sampled, expression-based clustering will allow for unbiased cell types reconstruction without previous knowledge of marker genes that define a cell type. Single-cell RNA sequencing is therefore providing powerful means to fully understand gene transcription regulatory network dynamics during development or pathological process. As the technologies continue improving, we anticipate that in the near future, numerous single-cell transcriptome datasets will be collected and a data driven biological and medical study era will eventually come.

References

Anders S, Huber W. Differential expression analysis for sequence count data. Genome Biol. 2010;11:R106. doi:10.1186/gb-2010-11-10-r106.

Batish M, Raj A, Tyagi S. Single molecule imaging of RNA in situ. Methods Mol Biol. 2011;714:3–13. doi:10.1007/978-1-61779-005-8_1.

Bendall SC, Simonds EF, Qiu P, et al. Single-cell mass cytometry of differential immune and drug responses across a human hematopoietic continuum. Science. 2011;332:687–96. doi:10.1126/science.1198704.

Bontoux N, Dauphinot L, Vitalis T, et al. Integrating whole transcriptome assays on a lab-on-a-chip for single cell gene profiling. Lab Chip. 2008;8:443–50. doi:10.1039/b716543a.

Bota M, Dong H-W, Swanson LW. From gene networks to brain networks. Nat Neurosci. 2003;6:795–9. doi:10.1038/nn1096.

Bruus H. Theoretical microfluidics. Oxford: Oxford University Press; 2007. ISBN 10: 0199235090/ 13: 9780199235094.

Bullard JH, Purdom E, Hansen KD, Dudoit S. Evaluation of statistical methods for normalization and differential expression in mRNA-seq experiments. BMC Bioinformatics. 2010;11:94. doi:10.1186/1471-2105-11-94.

Cann GM, Gulzar ZG, Cooper S, et al. mRNA-seq of single prostate cancer circulating tumor cells reveals recapitulation of gene expression and pathways found in prostate cancer. PLoS ONE. 2012;7:e49144. doi:10.1371/journal.pone.0049144.

Casbon JA, Osborne RJ, Brenner S, Lichtenstein CP. A method for counting PCR template molecules with application to next-generation sequencing. Nucleic Acids Res. 2011;39:e81. doi:10.1093/nar/gkr217.

Chao T-C, Ros A. Microfluidic single-cell analysis of intracellular compounds. J R Soc Interface. 2008;5 Suppl 2:S139–50. doi:10.1098/rsif.2008.0233.focus.

Di Carlo D, Lee LP. Dynamic signal-cell analysis for quantitative biology. Anal Chem. 2006;78:7918–25. doi:10.1021/ac069490p.

Dobson AT, Raja R, Abeyta MJ, et al. The unique transcriptome through day 3 of human preimplantation development. Hum Mol Genet. 2004;13:1461–70. doi:10.1093/hmg/ddh157.

Esumi S, Wu SX, Yanagawa Y, et al. Method for single-cell microarray analysis and application to gene-expression profiling of GABAergic neuron progenitors. Neurosci Res. 2008;60:439–51. doi:10.1016/j.neures.2007.12.011.

Evans E, Yeung A. Apparent viscosity and cortical tension of blood granulocytes determined by micropipet aspiration. Biophys J. 1989;56:151–60. doi:10.1016/S0006-3495(89)82660-8.

Fairfield FR, Newport JW, Dolejsi MK, von Hippel PH. On the processivity of DNA replication. J Biomol Struct Dyn. 1983;1:715–27. doi:10.1080/07391102.1983.10507477.

Fan R, Vermesh O, Srivastava A, et al. Integrated barcode chips for rapid, multiplexed analysis of proteins in microliter quantities of blood. Nat Biotechnol. 2008;26:1373–8. doi:10.1038/nbt.1507.

Fan HC, Wang J, Potanina A, Quake SR. Whole-genome molecular haplotyping of single cells. Nat Biotechnol. 2011;29:51–7. doi:10.1038/nbt.1739.

Fu GK, Hu J, Wang P-H, Fodor SPA. Counting individual DNA molecules by the stochastic attachment of diverse labels. Proc Natl Acad Sci U S A. 2011;108:9026–31. doi:10.1073/pnas.1017621108.

Gaynor EM, Mirsky ML, Lewin HA. Use of flow cytometry and RT-PCR for detecting gene expression by single cells. Biotechniques. 1996;21:286–91.

Hashimshony T, Wagner F, Sher N, Yanai I. CEL-seq: single-cell RNA-seq by multiplexed linear amplification. Cell Rep. 2012;2:666–73. doi:10.1016/j.celrep.2012.08.003.

Hebenstreit D. Methods, challenges and potentials of single cell RNA-seq. Biology (Basel). 2012;1:658–67. doi:10.3390/biology1030658.

Islam S, Kjällquist U, Moliner A, et al. Characterization of the single-cell transcriptional landscape by highly multiplex RNA-seq. Genome Res. 2011;21:1160–7. doi:10.1101/gr.110882.110.

Islam S, Kjällquist U, Moliner A, et al. Highly multiplexed and strand-specific single-cell RNA 5′ end sequencing. Nat Protoc. 2012;7:813–28. doi:10.1038/nprot.2012.022.

Islam S, Zeisel A, Joost S, et al. Quantitative single-cell RNA-seq with unique molecular identifiers. Nat Methods. 2014;11:163–6. doi:10.1038/nmeth.2772.

Ivanov D, Dvoriantchikova G, Barakat DJ, et al. Differential gene expression profiling of large and small retinal ganglion cells. J Neurosci Methods. 2008;174:10–7. doi:10.1016/j.jneumeth.2008.06.016.

Jiang L, Schlesinger F, Davis CA, et al. Synthetic spike-in standards for RNA-seq experiments. Genome Res. 2011;21:1543–51. doi:10.1101/gr.121095.111.
Joo KM, Nam D-H. Prospective identification of cancer stem cells with the surface antigen CD133. Methods Mol Biol. 2009;568:57–71. doi:10.1007/978-1-59745-280-9_5.
Kaajakari V. MEMS tutorial: pull-in voltage in electrostatic microactuators. Practical MEMS: Design of microsystems, accelerometers, gyroscopes, RF MEMS, optical MEMS, and microfluidic systems. Small Gear Publishing; 2009, p. 1–2. ISBN-10: 0982299109, ISBN-13: 978–0982299104.
Kastrinaki MC, Andreakou I, Charbord P, Papadaki HA. Isolation of human bone marrow mesenchymal stem cells using different membrane markers: comparison of colony/cloning efficiency, differentiation potential, and molecular profile. Tissue Eng Part C Methods. 2008;14:333–9.
Kauzlarick D. Fundamentals of microfabrication, the science of miniaturization. 2nd ed [Book review]. IEEE Eng Med Biol Mag. 2003;22(2):109–111. doi:10.1109/MEMB.2003.1195708.
Kingston RE (2001) Preparation of poly(A) + RNA. Curr Protoc Mol Biol Chapter 4:Unit4.5. doi:10.1002/0471142727.mb0405s21
Kivioja T, Vähärautio A, Karlsson K, et al. Counting absolute numbers of molecules using unique molecular identifiers. Nat Methods. 2012;9:72–4. doi:10.1038/nmeth.1778.
Köster S, Angilè FE, Duan H, et al. Drop-based microfluidic devices for encapsulation of single cells. Lab Chip. 2008;8:1110–15. doi:10.1039/b802941e.
Krutzik PO, Nolan GP. Fluorescent cell barcoding in flow cytometry allows high-throughput drug screening and signaling profiling. Nat Methods. 2006;3:361–8. doi:10.1038/nmeth872.
Krutzik PO, Crane JM, Clutter MR, Nolan GP. High-content single-cell drug screening with phosphospecific flow cytometry. Nat Chem Biol. 2008;4:132–42. doi:10.1038/nchembio.2007.59.
Langmead B, Trapnell C, Pop M, Salzberg SL. Ultrafast and memory-efficient alignment of short DNA sequences to the human genome. Genome Biol. 2009;10:R25. doi:10.1186/gb-2009-10-3-r25.
Li J, Tibshirani R. Finding consistent patterns: a nonparametric approach for identifying differential expression in RNA-seq data. Stat Methods Med Res. 2011. doi:10.1177/0962280211428386.
Lovén J, Orlando DA, Sigova AA, et al. Revisiting global gene expression analysis. Cell. 2012;151:476–82. doi:10.1016/j.cell.2012.10.012.
Luo L, Callaway EM, Svoboda K. Genetic dissection of neural circuits. Neuron. 2008;57:634–60. doi:10.1016/j.neuron.2008.01.002.
Mamanova L, Turner DJ. Low-bias, strand-specific transcriptome Illumina sequencing by on-flowcell reverse transcription (FRT-seq). Nat Protoc. 2011;6:1736–47. doi:10.1038/nprot.2011.399.
Mamanova L, Andrews RM, James KD, et al. FRT-seq: amplification-free, strand-specific transcriptome sequencing. Nat Methods. 2010;7:130–2. doi:10.1038/nmeth.1417.
Manz A, Graber N, Widmer HM. Miniaturized total chemical analysis systems: a novel concept for chemical sensing. Sensors Actuators B Chem. 1990;1:244–8. doi:10.1016/0925-4005(90)80209-I.
Marcus JS, Anderson WF, Quake SR. Microfluidic single-cell mRNA isolation and analysis. Anal Chem. 2006;78:3084–9. doi:10.1021/ac0519460.
Marzluff WF, Wagner EJ, Duronio RJ. Metabolism and regulation of canonical histone mRNAs: life without a poly(A) tail. Nat Rev Genet. 2008;9:843–54. doi:10.1038/nrg2438.
McKenna BK, Evans JG, Cheung MC, Ehrlich DJ. A parallel microfluidic flow cytometer for high-content screening. Nat Methods. 2011;8:401–3. doi:10.1038/nmeth.1595.
Nguyen NT, Wereley S (2006) Fundamentals and applications of microfluidics. Design 471
Novick A, Weiner M. Enzyme induction as an all-or-none phenomenon. Proc Natl Acad Sci U S A. 1957;43:553–66. doi:10.1073/pnas.43.7.553.
Ortín J, Parra F. Structure and function of RNA replication. Annu Rev Microbiol. 2006;60:305–26. doi:10.1146/annurev.micro.60.080805.142248.

Ozsolak F, Milos PM. RNA sequencing: advances, challenges and opportunities. Nat Rev Genet. 2011;12:87–98. doi:10.1038/nrg2934.
Perfetto SP, Chattopadhyay PK, Roederer M. Seventeen-colour flow cytometry: unravelling the immune system. Nat Rev Immunol. 2004;4:648–55. doi:10.1038/nri1416.
Picelli S, Björklund ÅK, Faridani OR, et al. Smart-seq2 for sensitive full-length transcriptome profiling in single cells. Nat Methods. 2013;10:1096–8. doi:10.1038/nmeth.2639.
Qiu S, Luo S, Evgrafov O, Li R. Single-neuron RNA-seq: technical feasibility and reproducibility. Front Genet. 2012;3:124.
Ramsköld D, Luo S, Wang Y-C, et al. Full-length mRNA-seq from single-cell levels of RNA and individual circulating tumor cells. Nat Biotechnol. 2012;30:777–82. doi:10.1038/nbt.2282.
Robinson MD, Oshlack A. A scaling normalization method for differential expression analysis of RNA-seq data. Genome Biol. 2010;11:R25. doi:10.1186/gb-2010-11-3-r25.
Sasagawa Y, Nikaido I, Hayashi T, et al. Quartz-seq: a highly reproducible and sensitive single-cell RNA sequencing method, reveals non-genetic gene-expression heterogeneity. Genome Biol. 2013;14:R31. doi:10.1186/gb-2013-14-4-r31.
Schmid A, Kortmann H, Dittrich PS, Blank LM. Chemical and biological single cell analysis. Curr Opin Biotechnol. 2010;21:12–20. doi:10.1016/j.copbio.2010.01.007.
Shalek AK, Satija R, Adiconis X, et al. Single-cell transcriptomics reveals bimodality in expression and splicing in immune cells. Nature. 2013;498:236–40. doi:10.1038/nature12172.
Shiroguchi K, Jia TZ, Sims PA, Xie XS. Digital RNA sequencing minimizes sequence-dependent bias and amplification noise with optimized single-molecule barcodes. Proc Natl Acad Sci U S A. 2012;109:1347–52. doi:10.1073/pnas.1118018109.
Spruce AE, Iwata A, White JM, Almers W. Patch clamp studies of single cell-fusion events mediated by a viral fusion protein. Nature. 1989;342:555–8. doi:10.1038/342555a0.
Sultan M, Schulz MH, Richard H, et al. A global view of gene activity and alternative splicing by deep sequencing of the human transcriptome. Science. 2008;321:956–60. doi:10.1126/science.1160342.
Tang F, Barbacioru C, Wang Y, Nordman E. mRNA-seq whole-transcriptome analysis of a single cell. Nat Methods. 2009;6(5):377–82. doi:10.1038/NMETH.1315.
Tariq MA, Kim HJ, Jejelowo O, Pourmand N. Whole-transcriptome RNAseq analysis from minute amount of total RNA. Nucleic Acids Res. 2011;39:e120. doi:10.1093/nar/gkr547.
Trapnell C, Pachter L, Salzberg SL. TopHat: discovering splice junctions with RNA-seq. Bioinformatics. 2009;25:1105–11. doi:10.1093/bioinformatics/btp120.
Trapnell C, Williams BA, Pertea G, et al. Transcript assembly and quantification by RNA-seq reveals unannotated transcripts and isoform switching during cell differentiation. Nat Biotechnol. 2010;28:511–15. doi:10.1038/nbt.1621.
Trapnell C, Roberts A, Goff L, et al. Differential gene and transcript expression analysis of RNA-seq experiments with TopHat and Cufflinks. Nat Protoc. 2012;7:562–78. doi:10.1038/nprot.2012.016.
Treutlein B, Brownfield DG, Wu AR, et al. Reconstructing lineage hierarchies of the distal lung epithelium using single-cell RNA-seq. Nature. 2014;509(7500):371–5. doi:10.1038/nature13173.
Wheeler AR, Throndset WR, Whelan RJ, et al. Microfluidic device for single-cell analysis. Anal Chem. 2003;75:3581–6. doi:10.1021/ac0340758.
White AK, Vaninsberghe M, Petriv OI, et al. High-throughput microfluidic single-cell RT-qPCR. Proc Natl Acad Sci U S A. 2011;108:13999–4004. doi:10.1073/pnas.1019446108.
Wu AR, Neff NF, Kalisky T, et al. Quantitative assessment of single-cell RNA-sequencing methods. Nat Methods. 2013;11(1):41–6. doi:10.1038/nmeth.2694.
Yan L, Yang M, Guo H, et al. Single-cell RNA-seq profiling of human preimplantation embryos and embryonic stem cells. Nat Struct Mol Biol. 2013;20:1131–9. doi:10.1038/nsmb.2660.
Zhu YY, Machleder EM, Chenchik A, et al. Reverse transcriptase template switching: a SMART approach for full-length cDNA library construction. Biotechniques. 2001;30:892–7.

Kenian Chen In 2004, Dr. Kenian Chen received BS degree in biotechnology, followed by PhD degree in biochemistry and molecular biology from Sun Yat-Sen University in 2009. He has ten years of experience in Bioinformatics, and authored a number of publications involved in molecular evolution and systems biology. From 2009.7 to 2011.11, Kenian Chen worked as a lecturer in Department of Biotechnology, Guangzhou Medical University, teaching biochemistry and bioinformatics for undergraduate students. Since 2011.11, Kenian Chen worked in Prof. Jiaqian Wu's lab as a Postdoc fellow. He focuses on data analyses generated from next generation sequencing, including single cell sequencing, in order to understand stem cell differentiation and molecular mechanism of CNS development and neurological diseases.

Xiaomin Dong Dr. Xiaomin Dong obtianed his Ph.D. degree in the Department of Gene Therapy, Ulm University, Germany in 2012, with his thesis on the application of viral vectors in the research of neurodegenerative disorders. From 2012 to 2013, he worked in a research team as the group leader to study focal cerebral ischemic stroke in the Model Animal Research Center of Wuhan University, China by using transgenic and knockout mouse models. In 2013, he joined Prof. Jiaqian Wu's lab in the Vivian L. Smith Department of Neurosurgery at the University of Texas as a postdoctoral fellow. His main research interests focus on the role of regulatory RNAs in neural differentiation.

Jiaqian Wu An assistant professor in the Vivian L. Smith Department of Neurosurgery and Center for Stem Cell and Regenerative Medicine at the University of Texas Medical School at Houston, Dr. Wu earned her doctorate in molecular and human genetics at Baylor College of Medicine and did her postdoctoral work at Yale and Stanford University. Wu laboratory combines stem cell biology and systems-based approaches involving functional genomics, bioinformatics and next-generation sequencing technologies to unravel gene transcription and regulatory mechanisms governing neural and blood development and differentiation. Dr. Wu's work has been recognized with prestige honors and awards, including the National Institute of Health Pathway to Independence (PI) Award (K99/R00) and the Senator Lloyd & B.A. Bentsen Investigator Award which she currently holds; the National Institutes of Health Ruth L. Kirschstein National Research Service Award for Individual Postdoctoral Fellows; and the International Society for Stem Cell Research (ISSCR) Annual Meeting Travel Award etc. A reviewer for NIH, the journals Nucleic Acids Research, Genome Research, and Genome Biology, Dr. Wu has presented invited talks and lectures on stem cell biology, functional genomics, and proteomics at international conferences, the Multiple Sclerosis Research Center of New York, Lawrence Livermore National Laboratory, and the University of Florida etc. She has developed a patent, authored a book, and wrote many articles that have appeared in PNAS, Genome Biology, Plos Genetics, Genome Research, BioTechniques, and Nature, among others.

Chapter 5
System Immunology in Metabolic Research of Dendritic Cells

Duojiao Wu

Abstract Dendritic cells (DCs) are key regulators of both immunity and tolerance by controlling activation and polarization of effector T helper cell and regulatory T cell responses. It is well known that cellular activation and function are coupled to cellular metabolism. There is a growing appreciation that these metabolic changes also underlie the capacity of immune cells to perform particular functions. This has led to the concept that the manipulation of cellular metabolism can be used to shape innate and adaptive immune responses. However, it is difficult to explore metabolic underpinnings of immunity across the entire spectrum of innate and adaptive immune responses. Systems immunology aims to give a synthetic understanding of how these components function together as a whole instead of analyzing individual components of immune systems. With an increasing interest in systems-level studies spawned by high-throughput technologies, many immunologists are exploring whether targeting of metabolic pathways in DCs could be a novel approach to manipulate the functional properties of DCs for clinical purposes.

Keywords Systems immunology • Metabolic • Dendritic cells • Sequencing • Inflammation

Over decades, the field of immunology has been growing at an exponential pace. Today, there is an increase in studying how immune cells function in terms of their intracellular metabolism (van der Windt et al. 2012), how these metabolic pathways affect the phenotype and activation of immune cells (Ferrante 2013), and how the immune system affects the metabolic functions of its host organism also known as "immunometabolism" (Pearce and Pearce 2013; Granger and Emambokus 2013). However, the present data do not take into account the observed diversity in immune cell lineages or their tissue-specific functions. Immune systems have a high degree of interdependence and interconnection between components. We are limited in

D. Wu, M.D., Ph.D. (✉)
Biomedical Research Center, Zhongshan Hospital, Fudan University, Shanghai, China

Shanghai Institute of Clinical Bioinformatics, Shanghai, China
e-mail: wu.duojiao@zs-hospital.sh.cn

X. Wang (ed.), *Single Cell Sequencing and Systems Immunology*,
Translational Bioinformatics 5, DOI 10.1007/978-94-017-9753-5_5

our understanding of how the immune system operates when we focus only on the properties or actions of one or a few unconnected components.

Systems immunology is an emerging discipline that combines high-content, multiplexed measurements with informatic and computational modeling methods to better understand immunological function at various scales (Narang et al. 2012). Here we provide descriptions of the key data gathering techniques employed to generate the quantitative and qualitative data and summarize the progress to date in applying these tools and techniques to questions of immunological interest. Dendritic cells (DCs), which are present in nearly all tissues of the body, continually sample their external environment and are critical for initiating antigen-specific adaptive immune responses and for promoting tolerance (Everts and Pearce 2014). The present chapter will take metabolic regulation of DCs as an example and illustrate how to use the tools of system immunology to eventually produce a better understanding of immune responses.

5.1 Introduction of System Immunology

Immunological studies frequently analyze individual components (e.g., signaling pathways) of immune systems in a reductionist manner. Recently, the need to have an integrated understanding of how these individual components work together in a system is being increasingly realized. System immunology is developed based on the improved technical capacity as following: (1) the widespread use of deep sequencing methods; (2) the advent of highly multiplexed flow cytometry; (3) the availability of high-throughput proteomics; (4) the development of bioinformatics, which aims to give a synthetic understanding of how these components function together as a whole (Fig. 5.1).

To achieve the integrative understanding, systems immunology complements empirical and experimental approaches with modeling and simulation. A model represents a real system which synthesizes existing knowledge of multiscale and transcale integration (population, organism, organ, tissue, cell, etc.). In a model, each factor correspond to real component of real system and computational algorithm governing the model correspond to real physical laws. A model can also serve to generate new knowledge about the system by simulation. Simulation refers to operating a model under a configuration of interest to "simulate" the system's behavior.

For instance, immunologists have long recognized the importance of early receptor-receptor ligation events in regulating cell function and differentiation state. However, so many molecules and proteins are involved in the events. It is hard to put pieces together. Also there is increasing evidence for a crucial role of the biochemical and conformational properties of the membrane and proximal act in and myosin fiber structures in guiding such events during lymphocyte interactions (Delon et al. 2001; Burkhardt et al. 2008). These insights have prompted numerous theoretical studies focusing on the small-scale spatial aspects of cellular signaling.

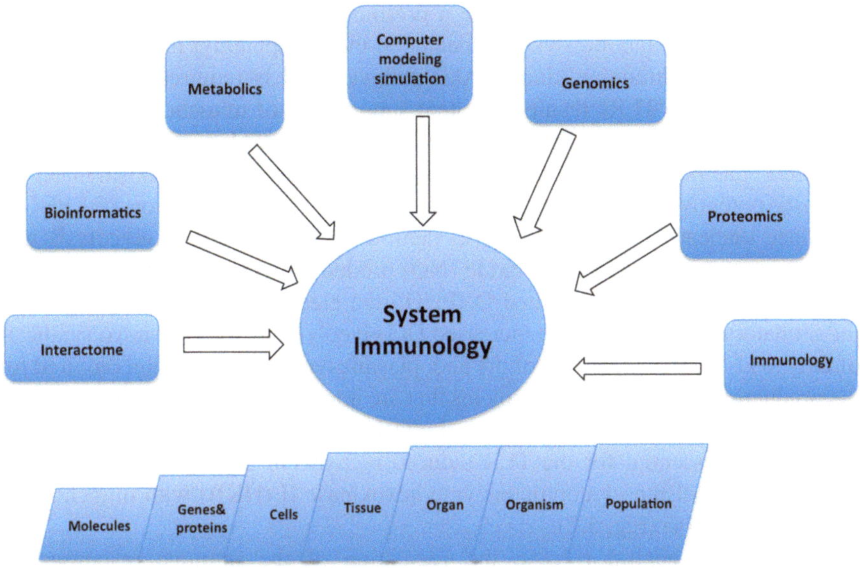

Fig. 5.1 Development of system immunology
The figure illustrates system biology is developed based on the systems-level studies spawned by high-throughput technologies with a focus on dynamic networks and computer modeling/simulation processes

Stochastic spatially (Andrews and Bray 2004) resolved simulations explore the early kinetics of signaling processes within small networks of individually interacting proteins and lipids. Using an intuitive picture of chemical reaction systems, each molecule is treated as a point-like particle that diffuses freely in three-dimensional space. A reaction occurs and the simulated reactants are replaced by products. Many of these approach utilize Monte Carlo2 methods (Horn and Lackner 2014; Cala et al. 2011) to incorporate the random thermal fluctuations governing molecular Brownian motion and reactive encounters.

5.2 Progress in Metabolic Regulation of Dendritic Cells

Tissue-resident DCs exhibit great heterogeneity in gene and cell-surface marker expression, which reflects their distinct capabilities for antigen processing and engagement of effector lymphocytes. Therefore, their bioenergetic demands are likely to be distinct and related to their tissue-specific functions (Ganeshan and Chawla 2014). Currently, the studies of how metabolic substrates and pathways fuel DCs are primarily derived from studies performed using bone marrow–derived DCs. The maturation of DCs is directed by signal transduction events downstream of toll-like receptors (TLRs) and other pattern recognition receptors. Krawczyk et al.

(2010) found TLR agonists stimulate a profound metabolic transition to aerobic glycolysis. This metabolic switch depends on the phosphatidyl inositol 3′-kinase (PI3k)/the serine/threonine kinase (Akt) pathway, and is required for DC maturation. Ligation of TLRs results in potent suppression of mitochondrial oxidative metabolism, suggesting that this metabolic checkpoint needs to be cleared for the complete enactment of the activated phenotype. However, the mechanism that TLR agonists potently suppress mitochondrial oxidative phosphorylation (OXPHOS) remained unclear. Everts et al. (2012) demonstrated this shift to glycolysis was associated with increased nitric oxide (NO) production via nitric oxide synthase 2 (NOS2), which potently inhibited OXPHOS the electron transport chain in these inflammatory DCs. Along with increased DC glycolytic flux, the de novo synthesis of fatty acids for the expansion of the endoplasmic reticulum and Golgi required for the production and secretion of proteins that are integral to DC activation. Signaling via the kinases TANK-binding kinase 1 (TBK1), IkappaB kinase epsilon (IKKε) and Akt was essential for the TLR-induced increase in glycolysis by promoting the association of the glycolytic enzyme hexokinase II (HKII) with mitochondria (Everts et al. 2014) (Fig. 5.2).

DCs play a central role in the regulation of the immune responses by providing the information needed to decide between tolerance, ignorance, or active

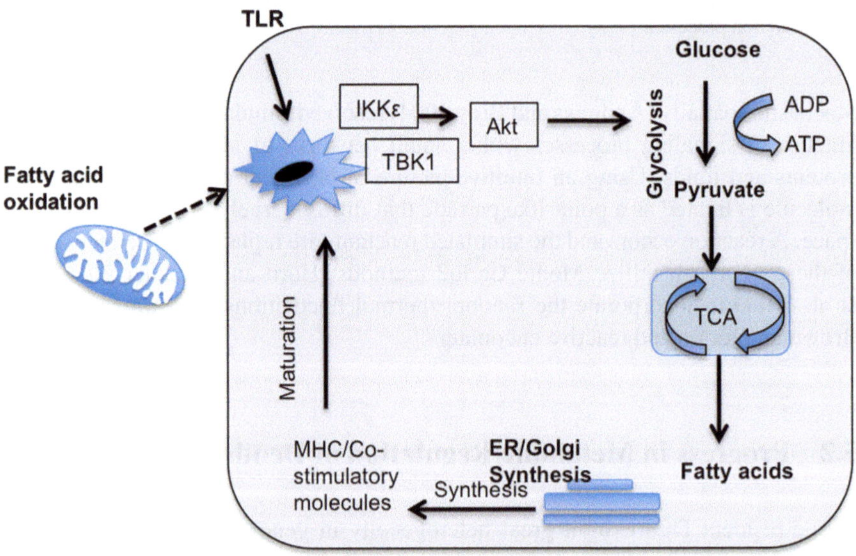

Fig. 5.2 Dendritic cells make switch to glycolysis with the stimulation of TLR agonists
In response to TLR ligands, resting DCs undergo maturation, resulting in their migration to lymphoid tissues where processed antigens are presented to naive T cells. The ligation of TLRs in DCs activates the metabolic switch for glycolytic metabolism, a program that is regulated via the kinases TBK1, IKKε and Akt pathway

responses. For this reason different therapies aim at manipulating DC to obtain the desired response, such as enhanced cell-mediated toxicity against tumor and infected cells or the induction of tolerance in autoimmunity and transplantation (Seliger and Massa 2013). Tolerogenic DCs, as opposed to immunogenic DCs, are generally characterized by the absence of traditional signs of activation, are maturation-resistant, and express increased levels of immunoregulatory factors, important for controlling the regulatory T cell (Treg) responses (Kornete and Piccirillo 2012). Recent findings now show that increased accumulation of lipids in DCs in tumors can impair presentation of tumor antigens-which is crucial for activation of the immune system-providing a conceptual framework that may help increase the efficacy of therapeutic vaccines (Zitvogel and Kroemer 2010; Herber et al. 2010).

Altogether, these studies suggest that the metabolic regulation of DCs contribute to the cellular maturation (some known and others unknown) and immune responses.

5.3 System Immunology in Metabolic Regulation of DCs

Despite rapid progress over the last decade, our understanding of immune cell metabolism is still far from enough. First of all, the understanding of metabolic effector responses across the entire repertoire of immune cells is fractional. Most of investigations should come from tissue-resident or recruited cells that actively participate in immune defense and tissue homeostasis. Meanwhile, the present data do not consider the observed diversity in immune cell lineages or their tissue-specific functions. Second, we know little about how these recognized signals are integrated into a cohesive metabolic program, which supports the cells' effector functions. Thus, it will be important to develop novel tools and experimental models to explore metabolic underpinnings of immunity across the entire spectrum of innate and adaptive immune responses.

Today the advances in genetics and proteomics promise to usher in a new era of immunometabolism research. Given the host defense and tissue homeostasis is complicated, a disease is rarely a consequence of an abnormality in a single gene, but reflects the perturbations of the complex intracellular and intercellular network that links tissue and organ systems (Amato et al. 2014). The emerging tools of system immunology offer a platform to explore systematically not only the molecular complexity of a particular disease, leading to the identification of disease modules and pathways, but also the molecular relationships among distinct phenotypes as a whole.

In recent years, people started to apply large-scale data acquisition technologies into DCs research. By using iTRAQ labeling, Ferret-Bernard et al. (2012) found the molecules up-regulated in pro-Th2 DC compared to pro-Th1 DCs including some potentially involved in protein folding during antigen processing also enzymes/transporters related to cell metabolism (malate dehydrogenase, pyruvate kinase). The study suggests that pro-Th2 DCs are more metabolically active while pro-Th1 DCs have a mature 'end state' (Ferret-Bernard et al. 2012). Fabrik et al. (2014) found DCs have a very active arginine metabolism in the progressed infection. Metabolic regulation is also involved in the differentiation of DC progenitors (Luber et al. 2010).

Computer modeling/simulation is also employed to study "immunometabolism". The immune cells likely optimize flux through various metabolic pathways in a dynamic manner to fuel their emerging bioenergetic needs. It is very delicate regulations which cannot simply be analyzed as mitochondria-based oxidative metabolism or glycolysis/glutaminolysis. Using computer modeling and simulations could simulate the kinetics of signaling processes within small networks of individually interacting proteins and lipids. Using an intuitive picture of chemical reaction systems, each molecule is treated as a point-like particle that diffuses freely in three-dimensional space.

We can summarize immune response as networks of many genes and proteins (Amato et al. 2014; Zhang et al. 2013). The nodes of network models represent concentrations or activation states of the different molecular species, whereas the links (or edges) encode interactions and state transitions. By using bioinformatics software, analyzing protein-protein interaction networks, scientists would easily find key regulation factors which is responsible for starting immune response (Barabási et al. 2011). Cellular decisions are determined by complex molecular interaction networks. Therefore, computational studies based upon the structure of these networks are of great interest. Networks have been applied to immunological systems such as in the analysis of TCR activation (Saez-Rodriguez et al. 2007) which are called Boolean networks because the rules for update determine the on- or off state for the next iteration, based on operations that produce 0 or 1 values and take as input logical combinations of 0 and 1 values.

5.4 Perspectives

The major goal of the metabolic rewiring is to provide the immune cell with sufficient energy (ATP) and metabolic intermediates to perform its effector functions in host defense and tissue homeostasis. Scientists believe that interference with or augmentation of certain metabolic programs might be clinically useful in dampening pathogenic autoimmunity or chronic inflammation in a diverse group of metabolic and degenerative diseases (Huang and Mellor 2014; Iyer et al. 2012). This is a futuristic view of system immunology—an integrated understanding of the interactions among the genome, the proteome, the environment and the pathophenome, mediated by the underlying cellular network, offers a basis for future advances. Such advances will help us to understand the structure and the workings of the wiring diagram—the prerequisite towards identifying the components whose functions need to be maintained and those whose activity must be altered with drugs.

References

Amato R, Morleo M, Giaquinto L, di Bernardo D, Franco B. A network-based approach to dissect the cilia/centrosome complex interactome. BMC Genomics. 2014;15:658.

Andrews SS, Bray D. Stochastic simulation of chemical reactions with spatial resolution and single molecule detail. Phys Biol. 2004;1:137–51.

Barabási AL, Gulbahce N, Loscalzo J. Network medicine: a network-based approach to human disease. Nat Rev Genet. 2011;12:56–68.

Burkhardt JK, Carrizosa E, Shaffer MH. The actin cytoskeleton in T cell activation. Annu Rev Immunol. 2008;26:233–59.

Cala O, Fabre S, Pinaud N, et al. Towards a molecular interpretation of astringency: synthesis, 3D structure, colloidal state, and human saliva protein recognition of procyanidins. Planta Med. 2011;77:1116–22.

Delon J, Kaibuchi K, Germain RN. Exclusion of CD43 from the immunological synapse is mediated by phosphorylation-regulated relocation of the cytoskeletal adaptor moesin. Immunity. 2001;15:691–701.

Everts B, Pearce EJ. Metabolic control of dendritic cell activation and function: recent advances and clinical implications. Front Immunol. 2014;5:203.

Everts B, Amiel E, van der Windt GJ, et al. Commitment to glycolysis sustains survival of NO-producing inflammatory dendritic cells. Blood. 2012;120:1422–31.

Everts B, Amiel E, Huang SC, et al. TLR-driven early glycolytic reprogramming via the kinases TBK1-IKKε supports the anabolic demands of dendritic cell activation. Nat Immunol. 2014;15:323–32.

Fabrik I, Link M, Härtlova A, Dankova V, Rehulka P, Stulik J. Application of SILAC labeling to primary bone marrow-derived dendritic cells reveals extensive GM-CSF-dependent arginine metabolism. J Proteome Res. 2014;13:752–62.

Ferrante AW. Macrophages, fat, and the emergence of immunometabolism. J Clin Invest. 2013;123:4992–3.

Ferret-Bernard S, Castro-Borges W, Dowle AA, et al. Plasma membrane proteomes of differentially matured dendritic cells identified by LC-MS/MS combined with iTRAQ labelling. J Proteomics. 2012;75:938–48.

Ganeshan K, Chawla A. Metabolic regulation of immune responses. Annu Rev Immunol. 2014;32:609–34.

Granger A, Emambokus N. Focus on immunometabolism. Cell Metab. 2013;17:807.

Herber DL, Cao W, Nefedova Y, et al. Lipid accumulation and dendritic cell dysfunction in cancer. Nat Med. 2010;16:880–6.

Horn H, Lackner S. Modeling of biofilm systems: a review. Adv Biochem Eng Biotechnol. 2014;145:189–214.

Huang L, Mellor AL. Metabolic control of tumour progression and antitumour immunity. Curr Opin Oncol. 2014;26:92–9.

Iyer A, Fairlie DP, Brown L. Lysine acetylation in obesity, diabetes and metabolic disease. Immunol Cell Biol. 2012;90:39–46.

Kornete M, Piccirillo CA. Functional crosstalk between dendritic cells and Foxp3(+) regulatory T cells in the maintenance of immune tolerance. Front Immunol. 2012;3:165.

Krawczyk CM, Holowka T, Sun J, et al. Toll-like receptor-induced changes in glycolytic metabolism regulate dendritic cell activation. Blood. 2010;115:4742–9.

Luber CA, Cox J, Lauterbach H, et al. Quantitative proteomics reveals subset-specific viral recognition in dendritic cells. Immunity. 2010;32:279–89.

Narang V, Decraene J, Wong SY, et al. Systems immunology: a survey of modeling formalisms, applications and simulation tools. Immunol Res. 2012;53:251–65.

Pearce EL, Pearce EJ. Metabolic pathways in immune cell activation and quiescence. Immunity. 2013;38:633–43.

Saez-Rodriguez J, Simeoni L, Lindquist JA, et al. A logical model provides insights into T cell receptor signaling. PLoS Comput Biol. 2007;3:e163.
Seliger B, Massa C. The dark side of dendritic cells: development and exploitation of tolerogenic activity that favor tumor outgrowth and immune escape. Front Immunol. 2013;4:419.
van der Windt GJ, Everts B, Chang CH, et al. Mitochondrial respiratory capacity is a critical regulator of CD8+ T cell memory development. Immunity. 2012;36:68–78.
Zhang B, Gaiteri C, Bodea LG, et al. Integrated systems approach identifies genetic nodes and networks in late-onset Alzheimer's disease. Cell. 2013;153:707–20.
Zitvogel L, Kroemer G. Targeting dendritic cell metabolism in cancer. Nat Med. 2010;16:858–9.

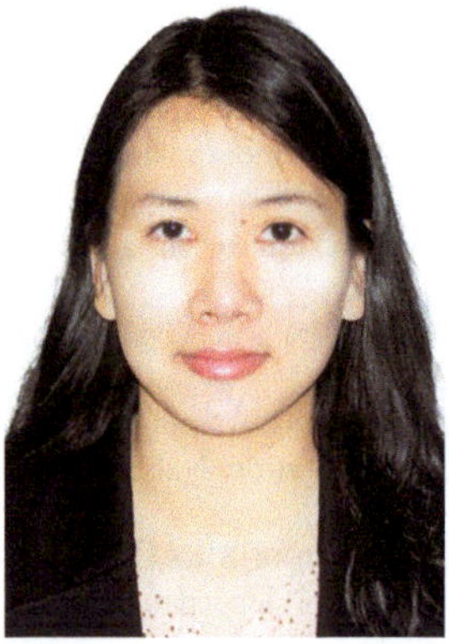

Duojiao Wu Associate Professor of Systems Immunology in the Biomedical Research Center of Fudan University Zhongshan Hospital. Principal Investigator and Director of Systems Immunology Group/Platform in Fudan University Center for Clinical Bioinformatics. She was selected and honored as a Shanghai Rising-star Scientist of Science and Technology Commission of Shanghai Municipality, China. Her research focuses on clinical bioinformatics, transplantation immunology, and systems immunology. Dr. Wu is funded by scientific grants including National Natural Science Foundation. She was invited to serve as reviewers of various international journals and as the faculty speaker in conferences. She has published more than 30 scientific papers in Journal of Translational Medicine, Journal of Proteome Research, and others.

Chapter 6
FOXP3$^+$ Treg Cells and Systems Biology Approaches to Studying Their Function

Xuerui Luo, Andy Tsun, Jing Yang, Zuojia Chen, Yayi Gao, Kongchen Li, Miranda Piccioni, and Bin Li

Abstract Regulatory T (Treg) cells play a key role in immunological homeostasis in health and disease. The molecular mechanisms underlying the differentiation, maintenance and function of Treg cells remains largely unclear. However, recent studies have shown that Treg development is not simply controlled by a series of linear signaling pathways, but rather a complex and dynamic regulatory network. This section provides an overview of the applications of systems biological tools in the Treg field. These approaches have brought to the field new insights into the complexities of Treg biology.

Keywords Regulatory T cells • FOXP3 • Systems biology • Metabolism

6.1 Introduction to Regulatory T (Treg) Cells

CD4$^+$ T cells are broadly characterized into two different types: effector T (Teff) and regulatory T (Treg) cells. Unlike Teff cells, Treg cells function as a suppressor of inflammatory effector responses to avoid misguided or excessive immune reactions. Therefore, Treg cells play an essential role in balancing immune responses and towards the prevention of autoimmunity (Fig. 6.1).

In human patients suffering from immunodysregulation polyendocrinopathy and enteropathy, X-linked syndrome (IPEX), several mutations in Forkhead box P3 (FOXP3 in humans, Foxp3 in mice – hereafter denoted as FOXP3), a member of the forkhead/winged-helix family of transcription factors, were identified as the cause for the development of the disease (Bennett et al. 2001). Concomitantly, in mice, the lack of functional FOXP3 was associated to the development of the 'Scurfy' phenotype, which is an X-linked recessive mutant. In these mice, disease progression is coupled with hyper-activated CD4$^+$ T cells, the overproduction of

X. Luo, B.S. • A. Tsun • J. Yang, B.S. • Z. Chen, Ph.D. • Y. Gao, Ph.D. • K. Li, M.S.
M. Piccioni, Ph.D. (✉) • B. Li, Ph.D. (✉)
Unit of Molecular Immunology, Key Laboratory of Molecular Virology and Immunology, Institut Pasteur of Shanghai, Chinese Academy of Sciences, 411 Hefei Road, Shanghai 200025, China
e-mail: pimiri@inwind.it; binli@sibs.ac.cn

X. Wang (ed.), *Single Cell Sequencing and Systems Immunology*,
Translational Bioinformatics 5, DOI 10.1007/978-94-017-9753-5_6

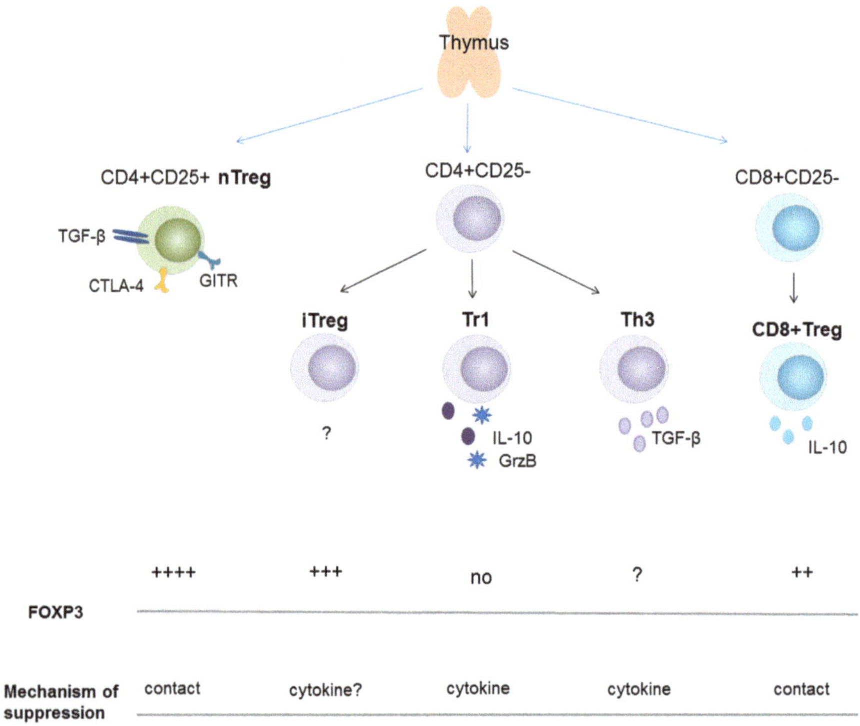

Fig. 6.1 Schematic representation of Treg Subsets differentiation and mechanisms of suppression

proinflammatory cytokines, and death within a month after birth (Wildin et al. 2001; Brunkow et al. 2001). Shortly after these findings, FOXP3 was found to be the master regulator of Treg cell differentiation and function. FOXP3 is expressed specifically in $CD4^{+}CD25^{+}$ Treg cells and determines their development and function in mitigating over-exuberant inflammatory responses (Hori et al. 2003; Fontenot et al. 2003).

As FOXP3 is important to Treg function, it is essential to understand the modulation of its expression and the post-translational modifications to which it may be subjected that affect its function. Studies of the FOXP3 promoter and its enhancers have revealed the transcription factors that are required to activate the expression of FOXP3 (Tone et al. 2008). DNA methylation has also been associated to FOXP3 expression particularly at the site dubbed the Treg-specific demethylated region (Floess et al. 2007; Liang et al. 2002). There are also FOXP3-specific miRNAs, such as miR-31, that targets the 3′-UTR of FOXP3 mRNA to mediate its down-regulation (Rouas et al. 2009).

FOXP3 can be acetylated to affect Treg suppressive function. Deacetylases, and signals from TGF-β and IL-6 have been shown to reduce the acetylation of FOXP3 to inhibit its binding to its target genes (Li et al. 2007; Samanta et al. 2008). Our lab has recently identified STUB1 as an E3 ligase of FOXP3. The upregulation of

STUB1 in Treg cells upon LPS stimulation or heat shock was found to promote FOXP3 degradation leading to the impairment of Treg function (Chen et al. 2013).

6.1.1 nTreg and iTreg Cells

Although Treg cells represent only 5–10 % of total $CD4^+$ T cells in peripheral blood mononuclear cells (PBMCs), they are involved in the prevention of autoimmune diseases, regulation of transplantation tolerance and the generation of tumors (Sakaguchi 2004). In general terms, Treg cells have been classified into two different subtypes, determined by their location of differentiation and development into mature Treg cells: one is the natural Treg (nTreg), the other the induced Treg (iTreg) (Table 6.1). nTreg cells are a thymus-derived subset of T cells that play a fundamental role in the maintenance of immune homeostasis, as demonstrated by the development of severe autoimmune diseases in new born mice (3–5 day) lacking a thymus. However, these symptoms can be relieved upon the injection of $CD4^+CD25^+$ T cells, showing that T cells that originate in the thymus can regulate and suppress self-destructive hyperactive immune responses (Asano et al. 1996).

Multiple factors have been shown to cooperate in the generation of nTreg cells in the thymus, such as TCR specificity, since nTreg development requires the presence of self-reactive TCRs and their respective cognate antigens (Jordan et al. 2001). The strength of the interaction between the TCR and MHC-peptide complexes, both in terms of affinity and avidity, directs thymocyte development towards a regulatory phenotype only when its threshold is below the level leading to negative selection and above the one driving positive selection (Feuerer et al. 2007; Atibalentja et al. 2009). The dramatic decrease of $CD25^+$ $FOXP3^-$ and $CD25^+$ $FOXP3^+$ T cells from CD28 and CD80/86 deficient mice has highlighted a fundamental role for co-stimulation, which could allow the escape of TCR-stimulated thymocytes from negative selection and favor the expression of FOXP3 (Hinterberger et al. 2011). Several works have revealed the importance of the NF-kB pathway in Treg development, with particular emphasis on PKC-θ, Bcl-10, CARMA1 and IKK2, whose defects in conditional knock-out mice heavily affect Treg cell generation (Feuerer et al. 2009). CD28-induced activation of c-Rel is considered a crucial event in priming developing thymocytes to a regulatory phenotype. Indeed c-Rel binds to

Table 6.1 The comparison between nTreg and iTreg cells

	nTreg	iTreg
Site of induction	Thymus	Peripheral lymphoid organs or inflammatory tissues
Mechanism of suppression	Cell contact dependent	Cytokine dependent
Specificity	Self-antigens (in thymus)	Foreign and self- antigens

a conserved non-coding DNA sequence (CNS-3) element at the *Foxp3* locus (Zheng et al. 2010) and also enhances the expression of the IL-2R (Vang et al. 2010). Finally, the abrogation of $FOXP3^{+}$ Treg cells concomitantly with the onset of autoimmune and inflammatory diseases in STAT5 deficient mice reveals the dependence on the IL-2R signaling pathway for proper Treg development (Burchill et al. 2007). Many other genes, factors and signaling pathways are thought to be involved and cooperate during Treg cell commitment, and the integration of different techniques is of primary importance towards the understanding of these complex networks.

Different from nTreg cells, iTreg cells are differentiated from naïve T cells in the periphery to acquire FOXP3 expression and suppressive function. This occurs through the activation of the TGF-β/IL-2 signaling pathways (Zheng et al. 2007; Kretschmer et al. 2005; Shevach et al. 2008), such as TGF-β-induced activation of Smad3 (Jana et al. 2009). There is a strong correlation between Treg and Th17 polarization, as the co-existence of TGF-β with other cytokines present in the microenvironment orchestrates the balance between these two subsets. Low concentrations of TGF-β, in the presence of IL-6, facilitates Th17 differentiation (Bettelli et al. 2006), while at high concentrations TGF-β cooperates with IL-2 to promote FOXP3 induction, inhibiting at the same time the expression of Th17-associated transcription factors (Fontenot et al. 2005; Zhou et al. 2008). Other classes of compounds, such as the metabolite of vitamin A, all-*trans* retinoic acid, can facilitate the differentiation of $FOXP3^{+}$ Treg cells and inhibit IL-6-driven induction of Th17 cells (Zhou et al. 2010).

FOXP3 expression alone is insufficient for the acquisition of a stable Treg phenotype as human naïve T cells can readily express FOXP3 upon TCR stimulation without acquiring any suppressive capacity (Gavin et al. 2006). As mentioned above, epigenetic modifications seem to guarantee the stability of regulatory T cells, as demonstrated by the increased demethylation status of the FOXP3 promoter observed in nTregs but not in TGF-β-induced Tregs (Floess et al. 2007). Lu and colleagues have highlighted that in TGF-β/atRA-iTreg cells, the acetylation of the H3 histone subunit plays a more predominant role than demethylation during iTreg differentiation (Lu et al. 2011).

Several groups claim that $CD4^{+}$ $CD25^{+}$ T cells isolated from PBMCs are a mixture of nTreg and iTreg cells (Horwitz et al. 2008), but controversial results and the lack of specific markers raise doubts over the real contribution of each subset to peripheral tolerance. Helios, a member of the Ikaros transcription factor family, was proposed to be selectively expressed by thymic-derived Treg cells (Thornton et al. 2010), but recent studies have reported its expression by TGF-β-induced Treg cells, at least under certain stimulatory conditions (Zabransky et al. 2012). The induction of neuropilin 1 (Nrp1) following ectopic expression of FOXP3 showed a strong linkage between Nrp1 and FOXP3 (Bruder et al. 2004). Recent studies have shown that Nrp1 may represent the most reliable marker for distinguishing nTreg from iTreg cells (Yadav et al. 2012; Weiss et al. 2012). Weiss and colleagues showed that $FOXP3^{+}$ Treg cells isolated from secondary lymphoid organs of wild-type mice are comprised of two distinct populations: $Nrp1^{high}$ and $Nrp1^{low}$ populations. Interestingly, in the first population the expression of Nrp1 is associated with high expression of Helios, while the second is characterized by the absence of Helios and high levels of death-associated protein-like 1 (DAPL-1) (Weiss et al. 2012). Thus, research on a combination of multiple molecules rather than a single chief marker seems to be the solution for distinguishing nTreg from iTreg cells, and further studies are required to validate this hypothesis and to reveal the role of these proteins during Treg cell development.

There are many proposed mechanisms by which Treg cells suppress immune responses, mostly characterized by the studying of nTreg cells. These include cytokine deprivation (IL-2), CTLA-4-mediated downregulation of CD80/CD86 co-stimulatory molecules on antigen presenting cells, and IL-35-mediated inhibition of effector T cell proliferation (Miyara and Sakaguchi 2007; Schmidt et al. 2012). However, it still remains unclear regarding the differences between nTreg and iTreg-mediated suppression, although recent studies have suggested both overlapping and separate roles of nTreg and iTreg cells in suppressing inflammation *in vivo* (Bilate and Lafaille 2012). Recent works have also revealed the expression in Treg cells of specific factors known to control the differentiation of other T cell lineages. GATA3, for a long time exclusively considered a canonical Th2-associated factor, was demonstrated to be essential for Treg in controlling immune homeostasis and inflammation, involved in FOXP3 expression and Treg accumulation at inflamed sites (Wang et al. 2011; Wohlfert et al. 2011), and shown to associate with FOXP3 to regulate the expression of different classes of genes, both in cooperative or antagonistic fashion (Rudra et al. 2012). A recent study from our group has also shed light on a molecular pathway that maintains GATA3 expression in nTreg cells. The E3 deubiquitinase USP21, whose expression is increased in a FOXP3-dependent manner, interacts with GATA3 promoting its stability via deubiquitination. In turn, GATA3 promotes FOXP3 function, thus creating a positive-feedback loop which may be fundamental in regulating Treg cell activity, especially under inflammatory conditions (Zhang et al. 2013). An extensive in depth analysis of FOXP3 complexes and their mechanisms of action would further our knowledge in this regard.

6.1.2 Other Treg Cells

Besides FOXP3$^+$ Treg cells, other Treg cell populations can be induced from naïve T cells in the periphery, including IL-10-producing type 1 Treg (Tr1 cells), TGF-β-producing Th3 cells and CD8$^+$ Treg cells.

6.1.2.1 Tr1 Cells

Adaptive T regulatory type 1 (Tr1) cells represent another subset of regulatory T cells inducible in the periphery from naïve CD4$^+$ T cells. The main characteristic that distinguishes them from other Treg populations is the absence of the constitutive expression of FOXP3, although they can transiently upregulate its expression

(Levings et al. 2005). *In vitro*, these cells can be induced form $CD4^+$ $CD45^+$ RO^- cells isolated from IPEX patients with different FOXP3 mutations. Although being characterized by undetectable levels of FOXP3 and reduced CD25 expression, induced-Tr1 cells display a low proliferative capacity following TCR stimulation and strong *in vitro* suppressive function towards allogeneic $CD4^+$ T cells. The development of suppressive Tr1 cells seems not to be dependent on FOXP3, given the presence of these cells in IPEX patients carrying $FOXP3^{null}$ mutations (Passerini et al. 2011). However, the serious illness characterizing these patients obviously reveals the inadequacy of Tr1 cells to compensate for the absence of nTreg cells. Tr1 cells are characterized by a well defined cytokine secretion profile, comprised of high levels of IL-10, TGF-β and IFN-γ *in vitro*, and low levels of IL-2 and IL-4 following antigen stimulation (Vieira et al. 2004).

Several markers have been proposed for Tr1 cells, such as lymphocyte-activation gene-3 (LAG-3), inducible-costimulatory molecule (ICOS) and programmed death receptor 1 (PD-1) (Haringer et al. 2009; Okamura et al. 2009; Akdis et al. 2004); however, these markers are also present on other types of IL-10 producing T cells and thus cannot be specifically used to identify Tr1 cells. One study also identified a $CD4^+$ $FOXP3^{low}$ $IL\text{-}10^+$ T cell population able to suppress autoimmunity and expressing CD49b (Charbonnier et al. 2006), but till now no progress has been made to further confirm CD49b as a marker for Tr1. However, Tr1 cells can be distinguished from other IL-10 producing T cell subsets because of their strict dependency on STAT-3 activation for the production of IL-10. IL-27 is actively involved in the differentiation of Tr1 cells, as it induces the phosphorylation of STAT-3 and the subsequent activation of Maf, a transcription factor involved both in the transactivation of the *il10* promoter and in the inhibition of FOXP3 expression (Pot et al. 2011). Recently, Tr1 cells that are able to suppress LPS-induced inflammatory responses have been generated in the presence of IL-6, which seem to participate in the induction of Tr1 from naïve $CD4^+$ T cells in the absence of IL-27 and TGF-β in a IL-21 and IL-2 dependent fashion (Jin et al. 2013).

The main suppressive mechanism exerted by Tr1 cells is through the secretion of large amounts of IL-10 and small quantities of TGF-β (Groux et al. 1997; Bacchetta et al. 1994), thus inhibiting the proliferation of effector T cells and the production of IL-2 and IFN-γ (Akdis and Blaser 2001), and influencing the differentiation and maturation of DCs and antigen presentation (Mosser and Zhang 2008). Besides cytokine secretion, Tr1 cells can suppress T cell responses in a cell-contact dependent mechanism, particularly through the expression of CTLA-4 and PD-1 (Akdis et al. 2004) and mediate granzyme-B dependent cytotoxic killing of myeloid APCs in an antigen-independent manner (Magnani et al. 2011).

6.1.2.2 Th3 Regulatory Cells

Another subset of suppressive T cells is represented by TGF-β-secreting Th3 cells, firstly identified in experimental autoimmune encephalomyelitis (EAE) mice and multiple sclerosis (MS) patients following oral tolerization with myelin basic protein (Fukaura et al. 1996; Chen et al. 1994). Th3 cells can only be triggered in an antigen-specific fashion but exert their suppressive activity in an antigen-non specific manner via the secretion of TGF-β. The characteristics of this lineage have been studied through the development of TGF/IL-$2^{-/-}$ transgenic mice, which express the intrinsic TGF-β transgene upon activation. Despite the lack of CD25, $CD4^+$ T cells from these transgenic mice display strong suppressive activity and high FOXP3 expression, and, differently from thymic-derived $CD25^+$ Tregs do not rely on IL-2 for their induction, differentiation and function, thus representing a separate regulatory T cell lineage (Carrier et al. 2007a). Th3 cells have also been demonstrated to contribute to peripheral tolerance by driving the differentiation of Ag-specific iTregs in the periphery, presumably due to the secretion of TGF-β (Carrier et al. 2007b). Moreover, by secreting TGF-β, Th3 regulatory T cells provide help for IgA secretion, and suppress both Th1 and Th2 responses (Fukaura et al. 1996). There are still controversial observations regarding the expression of FOXP3 by Th3 cells, and although some studies have shown that Th3 cells are characterized by an increase in FOXP3 expression and the up-regulation of CD25 and CTLA-4 (Carrier et al. 2007b; Ramsdell 2003; Jonuleit and Schmitt 2003), more in depth analysis is needed to unveil the possible markers, characteristics and their significance in the immune system.

6.1.2.3 $CD8^+$ Treg Cells

$CD8^+$ Treg cells are defined as regulatory cells due to their capacity to suppress autoimmunity and transplantation rejection (Mahic et al. 2008). Xystrakis and colleagues (2004) identified in the rat a $CD8^+$ $CD45RC^{low}$ population as cells producing IL-4, IL-10 and IL-13, and able to suppress the proliferation and differentiation of alloreactive $CD4^+$ T cells. These cells are not cytotoxic, display moderate levels of expression of FOXP3 and CTLA-4, and are able to protect rats from lethal GvHD (graft-versus-host disease). Natural $CD8^+$ $CD122^+$ regulatory T cells that are able to dampen autoimmunity after transfer into CD122-deficient mice have also been identified (Rifa'i et al. 2004), and the induction of $CD8^+$ regulatory T cells that suppress Th alloreactivity after pancreatic islet transplantation has been demonstrated (Vlad et al. 2008). $CD8^+$ Treg cells are also involved in the escape of immune responses by ocular tumors, which develop due to the expansion of $CD8^+$ Treg cells inducing apoptosis of responder T cells through a Fas/FasL dependent mechanism, TRAIL/DR5 interactions and possibly via TGF-β-dependent functional inactivation (McKenna and Previte 2012). $CD8^+$ Treg cells recognize antigen-activated $CD4^+$ T cells in a TCR-specific manner restricted by the MHC-class Ib molecule, Qa-1 (H2-T23; HLA-E in human), suggesting that they may be primed to TCR determinants expressed by $CD4^+$ T cells, and participate in the regulation of peripheral immunity during immune responses to both self and non-self antigens (Jiang and Chess 2000; Kim and Cantor 2011).

6.2 Application of Systems Biology Approaches in the Treg Field

Systems biology has emerged as a desirable approach for analyzing the workings of a cell on a broader scale compared to conventional methods. Through systematic genomic, proteomic, and metabolomic technologies, the construction of cell regulatory networks allows the acquisition of biological information at the whole cell level. Unlike other research approaches, systems biology treats all the aspects of a cell as one, and integrates various molecular clues to form a complete model. Several modeling approaches have been used for the study of Treg cells, such as microarray, high-throughput sequencing, mass spectroscopy and bioinformatics. Here is an overview of how systems biology has been used in the Treg field to answer key questions regarding their differentiation and function (Table 6.2).

6.2.1 Transcriptomics

The transcriptome contains the complete repertoire of RNA molecules produced by cells at any given time, which includes mRNA, rRNA, tRNA and other non-coding RNA. The transcriptome represents the functional elements on the genomic level, information that can be used to uncover the molecular mechanisms important for the development and function of T cells. Using transcriptomics, one can quantify the expression level of coding and non-coding transcripts, decipher the structure of genes and transcripts (5′- or 3′-UTR, splicing, and modifications), and detect the dynamic change in expression levels of these transcripts under different cellular environments and alterations in cellular signaling pathways.

Table 6.2 The application of systems biology approaches in the Treg field

Transcriptomics	Proteomics	Metabolism
• Microarry and ChIP	• Yeast two hybrid	• ^{1}H nuclear magnetic resonance (NMR) spectroscopy
• Next-generation sequencing	• Quantitative Mass spectrometry	• Partial least squares (PLS) methods
	• Tandem affinity purification	• Neural networks
	• Computational analysis	

6.2.1.1 Microarrays and Chromatin Immunoprecipitation (ChIP)

Microarray technology has been widely utilized to analyze RNA abundance within cells. This technique is based on the immobilization on a solid substrate of short oligonucleotide probes, complementary to the transcripts that have to be evaluated. The transcripts, previously extracted from tissue samples and labeled with fluorescent dyes, hybridize to the complementary probe on the array, which is then scanned with a laser. The resulting intensity of light is used as a surrogate of gene expression. Many studies have adopted microarrays to provide clues for further study in the Treg field.

Combined with chromatin immunoprecipitation (ChIP), microarray datasets provide a plethora of information to quantify the network of protein-DNA interactions. The first step in the ChiP-on-ChIP technology is represented by the cross-linking of the protein of interest to DNA, usually using formaldehyde. After cell lysis, the DNA is fragmented through sonication or other methods, and the protein-DNA fragments are immunoprecipitated. At this point, the protein of interest is no longer necessary for carrying on the procedure, and the complexes are gently dissociated. The DNA is fluorescently labeled, amplified and finally subjected to microarray.

Whole genomic microarrays and ChIP-on-Chip analysis have been used to analyze the differences in mRNA expression between nTreg vs. Teff cells, and to identify FOXP3-target genes in nTreg cells. Using this approach, several motifs for transcription factors known to play important roles in nTreg cells have been identified, including AP-1, Runx, NFAT and STAT proteins. Furthermore, the presence of FOXP3 binding sites close to loci encoding for 63 microRNAs has been identified, including miR-146a, miR-21, miR-155, miR-101 and miR-7, which are all potentially regulated by FOXP3 and may be fundamental for nTreg development and function (Sadlon et al. 2010).

Other studies have looked at the role of transcription factors other than FOXP3 in determining Treg development and function, such as SATB1, which is down-regulated in Treg cells. Using a similar approach, a recent paper demonstrated that the down-regulation of IL-22 observed in stimulated *ex vivo* Treg cells was associated with the direct binding of FOXP3 to the IL-22 promoter region (Jeron et al. 2012). As IL-22 is a Th17-associated cytokine, this data could be helpful for a better understanding of the role of FOXP3 in the plasticity of Treg cells. Microarray analysis of Treg cells transduced with SATB1 showed the reprogramming of nTreg cells into Teff cells and the induction of the Teff cell program (Beyer et al. 2011). Many of these studies have deposited their microarray expression files onto freely accessible internet databases. Given the expansion of the bioinformatics field, these databases should still be useful in the future for more in depth retrospective analysis.

6.2.1.2 Next-Generation Sequencing and Chromatin Immunoprecipitation (ChIP)

Next-generation sequencing (NGS) technologies are revolutionary genomic methods to sequence nucleotide sequences at high depth. Their use is becoming increasingly widespread following the fall in operational and material costs. An example is represented by the mRNA-Seq protocol, which was used to sequence between 12 and 29 million 32 bp cDNA fragments from different tissues, generating over 400 million reads in total and assessment of the expression of alternative transcript isoforms for different genes (Wang et al. 2008). This deep sequencing of small RNAs has also been used to determine miRNA abundance in mouse hematopoietic progenitor cells and downstream T cell lineages, aiming to reveal complex networks of immune microRNomes in developing lymphocytes (Kuchen et al. 2010).

NGS can also be coupled to ChIP. In the ChIP-Seq technique (Zhang et al. 2008; Mardis 2007), the immunoprecipitated DNA is massively sequenced to analyze the DNA fragments after ChIP, to gain insight into the relationship between transcription factor-DNA interactions. The advantage of these techniques resides in the possibility to study mRNA expression levels and even the differential expression of splice variants – the latter of which is a limitation of conventional microarrays that use defined DNA probes. The importance of this kind of approach in Treg cell-related research is represented by the study by Ouyang et al. on the role of Foxo1 in Treg function. Before this study, it was unclear as to whether Foxo1 acted independently or dependently on FOXP3 to modulate Treg cells. Using a genome-wide ChIP-sequencing approach, the authors found 300 Foxo1-bound target genes including previously unknown Foxo1-specific binding sites, showing an important and new role for Foxo1 in binding the *Ifng* gene to inhibit its expression (Ouyang et al. 2012). Interestingly, these genes were found to be regulated independent of FOXP3. Birzele et al. used the combination of ChIP-Seq and mRNA-Seq for the identification of FOXP3-binding regions and transcriptome profiling in resting and activated Teff and Treg cells in humans, detailing the differential expression of genes, splice variants and non-coding RNAs (Fig. 6.2) (Birzele et al. 2011). Future studies using a higher number of Seq-reads and more complex algorithms to analyze these data will no doubt uncover new relationships between gene expression and Treg function. Furthermore, the role of splice variation in defining cellular function has been of recent interest, as the detection of differential splice variants was limited by previous microarray methods.

6.2.2 Proteomics

The field of proteomics is the study of proteins on a large scale, where the protein snapshot is called the proteome. The main themes for exploration include the determination of quantitative protein levels, global protein-protein interactions, and post-translational modifications.

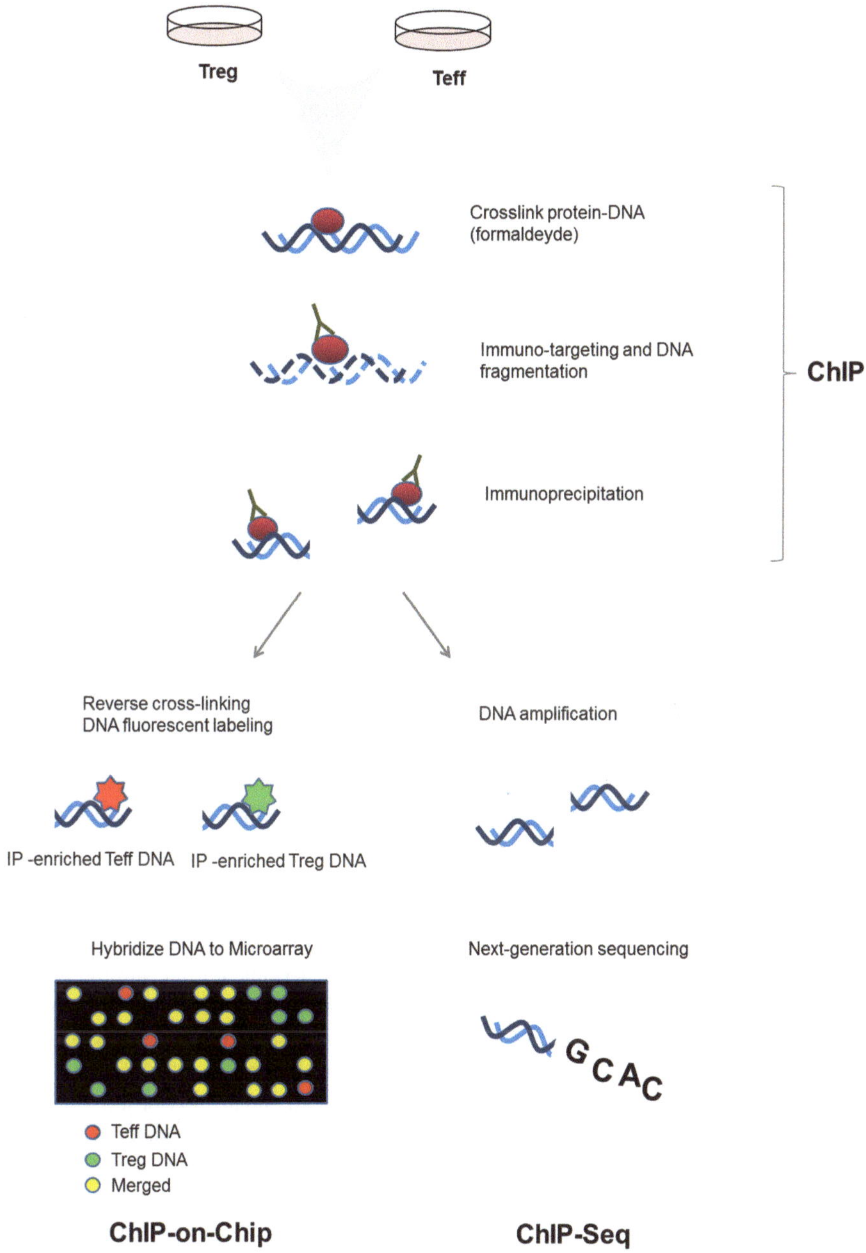

Fig. 6.2 Schematic representation of ChIP-on-Chip and ChIP-Seq workflows for the study of DNA-protein interactions

6.2.2.1 Yeast Two Hybrid System

The early adoption of proteomics was represented by the study of protein-protein interactions (PPIs) via the yeast two-hybrid system. Basically, interactions are revealed by mating two haploid yeast strains, one expressing a protein prey fused to a transcription activation domain (AD), and the other expressing a protein bait fused to a DNA-binding domain (DBD) that binds to a reporter gene. If the two proteins interact, the AD will be brought close to the DBD causing the transcription of the reporter gene (Ratushny and Golemis 2008). Several important FOXP3 binding partners have been identified using this method. For example, the interaction between FOXP3 and RUNX1, which was originally identified as a FOXP3 partner via a yeast two hybrid screen, is important for regulating Treg function by suppressing IL-2 expression, increasing the expression of Treg-associated molecules (CD25, GITR, CTLA-4) and enhancing Treg suppressive capacity (Ono et al. 2007). However, the limitation of yeast two hybrid screens includes its low accuracy and the artificial environment in which these binding studies take place. In particular, the occurrence of false positives has been estimated to account for about 50 % of the observed positive results, both due to technical artifacts, which lead to reporter activity even in the absence of interaction, and to biological PPIs that are not representative of the real cellular environment (Mrowka et al. 2001).

6.2.2.2 Mass Spectrometry

Mass spectrometry (MS) is a powerful tool to measure the protein content within a sample. From a technical point of view, mass spectrometry works by ionizing the peptide, for example through electrospray ionization (ESI) or matrix-assisted laser desorption/ionization (MALDI), and analyzing its mass-to-charge (m/z) ratio. In tandem MS or MS/MS, a full mass spectrum is recorded and then single peaks of interest can be selected, further fragmented and analyzed, providing information specific for the peptide amino-acid sequence. Post-translational modifications, including phosphorylation, acetylation, ubiquitination and methylation, can all be identified using this method. As specific modifications often occur in a small fraction of proteins in a sample, selective protein enrichment prior to MS analysis is often required. Enrichment can be achieved by chemical and affinity based methods, for example through the introduction of affinity tags allowing the selection of the modified protein or peptide (Angel et al. 2012). Using specific antibodies to capture the protein or post-translational modification of interest, researchers can analyze the changes in protein interactions or post-translational signals under different environmental contexts (Gronborg et al. 2002). Coupled with different techniques, mass spectrometry is a powerful tool for selectively analyzing specific protein or peptides. Tandem affinity purification (TAP) involves the creation of a fusion protein with a designed TAP tag. The protein complex of interest, such as FOXP3 complex, can be then purified through two-step affinity selection and the resulting interacting partners can be examined using mass spectrometry (Gao et al. 2014).

FOXP3 binds to DNA to regulate the transcriptional levels of its target genes. The binding capacity of FOXP3 is essential for Treg suppressive function and is assisted by FOXP3 cofactors, multimerization, and post-translational modification of FOXP3. An example of mass spectrometry application is shown by the work of Rudra et al. who purified FOXP3 and its complexes to explore their composition via mass spectrometry. They first generated a T cell line expressing a biotin-tagged FOXP3 to allow the purification of FOXP3-protein complexes using streptavidin magnetic bead chromatography. Purified protein complexes were analyzed by SDS-polyacrylamide gel electrophoresis (SDS-PAGE), protein bands were then excised, trypsin digested and subjected to micro-liquid chromatography-MS/MS. Their data confirmed that FOXP3 forms multiprotein complexes, and they identified 361 interactions, all of which may play a role in determining Treg differentiation and function through modulating the transcription of FOXP3 target genes (Rudra et al. 2012).

The development of mass spectrometry-based proteomics technologies and bioinformatic tools has also been used to discover novel post-translational modifications. Propionylation and butyrylation have been identified as new lysine modifications of histones (Basu et al. 2009; Chen et al. 2007). As such, there remains a huge area for exploring the differential characterization of the FOXP3 complex in Treg cells, including the various post-translational modifications of FOXP3, and its binding partners under different environmental stimuli.

6.2.2.3 Metabolomics

Metabolism in T cells is a highly dynamic process that regulates activation, differentiation, function and cell death (Gerriets and Rathmell 2012). T cell activation through TCR and CD28 stimulation initiates metabolic changes in order to support rapid cell growth, proliferation and the generation of effector cells. The metabolic pathways that are utilized are subtype-dependent upon activation. Th1, Th2, and Th17 cells decrease lipid oxidation and increase glycolysis, mainly through the up-regulation of glucose transporter 1 (Glut1), while regulatory T cells mostly rely on lipid oxidation (DeBerardinis et al. 2008; Michalek et al. 2011; Li et al. 2012). Hypoxia-inducible factor 1α (HIF1α) has been shown to skew metabolic signatures towards glycolysis and the differentiation of Th17 cells, away from Treg polarization. CD4^{+} T cells deficient in HIF1α fail to differentiate into IL-17 producing cells, as HIF1α is essential for driving T cell metabolism into the glycolytic pathway. Moreover, it can promote the transcription of Th17-associated genes, such as RORγt, by promoting FOXP3 degradation in a proteasome-dependent manner to attenuate Treg development (Dang et al. 2011; Shi et al. 2011; Tsun et al. 2011).

Leptin, a non-glycosylated protein classically considered as a hormone that regulates the balance between food intake and energy expenditure, is emerging as an immune response regulator, able to suppress Treg metabolism by regulating the cyclin-dependent kinase inhibitor 27 (p27kip1) and the phosphorylation of the extracellular-related kinases 1 (ERK1) and ERK2 (De Rosa et al. 2007). Other molecules, like Akt, mTOR, AMPK, and many others play pivotal roles in regulating the balance between Treg cells and other T cell subsets (Li et al. 2012). These studies show how metabolic factors may play roles other than those well characterized for metabolism. Due to the complexity of metabolic pathways, it is clear that the above high-throughput strategies may be utilized to uncover novel roles for metabolism in T cell differentiation.

6.2.2.4 Computational Analysis

Computational systems biology is used to analyze vast amounts of high-throughput data through the use of algorithms that infer and rank potential connections between target genes and possible regulators starting from large amounts of gene expression data-sets. Computational networks of biological systems have already been used to determine the transcription factors that make up the Treg cell signature (Fu et al. 2012). Researchers started from 129 gene-expression profiles previously obtained by microarray both from Teff and Treg cells. Using a bioinformatics platform, they selected 2021 potential transcription factors and 603 target genes that compose the known Treg signature. These data where then applied to an algorithm to predict Treg regulators. Beyond FOXP3 and factors already known, they identified many other additional candidates that may be involved in Treg function, both in activating and in a repressive fashion. With our increasing knowledge of protein structures and molecular pathways, we can now begin to predict or model putative molecular mechanisms and protein interactions.

6.3 Perspectives

In our immune system, Treg cells maintain immune homeostasis and prevent inflammatory diseases. Recently, a number of studies have suggested that Treg cells have the capacity to convert into other T cell types, a phenomenon known as Treg plasticity (Oldenhove et al. 2009; Rubtsov et al. 2010). However, a countering argument is that these converted cells are a minor population of uncommitted $FOXP3^{+}$ T cells (Hori 2010).

The development of high throughput single cell analysis may prove a significant advance in this field, as this would allow us to test for the exact expression signature that permits Treg plasticity without having to enrich any potential test populations. Advances in single cell analysis includes single molecule and cell bioimaging (Bokinsky et al. 2003; Rust et al. 2004), and also single molecule spectroscopy and nonlinearity optical imaging technology (Xie and Dunn 1994). A great deal of computational power would also have to supplement these advances in order to cope with increasingly large and complex datasets.

Another area that has not been covered is the potential of genomic RNA interference arrays to test the role of different genes in controlling Treg cell differentiation and function. The development of less toxic/harmful and simplified methods for introducing RNA into primary T cells would aid research in this direction.

References

Akdis CA, Blaser K. Mechanisms of interleukin-10-mediated immune suppression. Immunology. 2001;103(2):131–6.

Akdis M, et al. Immune responses in healthy and allergic individuals are characterized by a fine balance between allergen-specific T regulatory 1 and T helper 2 cells. J Exp Med. 2004;199(11):1567–75.

Angel TE, et al. Mass spectrometry-based proteomics: existing capabilities and future directions. Chem Soc Rev. 2012;41(10):3912–28.

Asano M, et al. Autoimmune disease as a consequence of developmental abnormality of a T cell subpopulation. J Exp Med. 1996;184(2):387–96.

Atibalentja DF, Byersdorfer CA, Unanue ER. Thymus-blood protein interactions are highly effective in negative selection and regulatory T cell induction. J Immunol. 2009;183(12): 7909–18.

Bacchetta R, et al. High levels of interleukin 10 production in vivo are associated with tolerance in SCID patients transplanted with HLA mismatched hematopoietic stem cells. J Exp Med. 1994;179(2):493–502.

Basu A, et al. Proteome-wide prediction of acetylation substrates. Proc Natl Acad Sci U S A. 2009;106(33):13785–90.

Bennett CL, et al. The immune dysregulation, polyendocrinopathy, enteropathy, X-linked syndrome (IPEX) is caused by mutations of FOXP3. Nat Genet. 2001;27(1):20–1.

Bettelli E, et al. Reciprocal developmental pathways for the generation of pathogenic effector T(H)17 and regulatory T cells. Nature. 2006;441(7090):235–8.

Beyer M, et al. Repression of the genome organizer SATB1 in regulatory T cells is required for suppressive function and inhibition of effector differentiation. Nat Immunol. 2011;12(9):898–907.

Bilate AM, Lafaille JJ. Induced CD4+ Foxp3+ regulatory T cells in immune tolerance. Annu Rev Immunol. 2012;30:733–58.

Birzele F, et al. Next-generation insights into regulatory T cells: expression profiling and FoxP3 occupancy in human. Nucleic Acids Res. 2011;39(18):7946–60.

Bokinsky G, et al. Single-molecule transition-state analysis of RNA folding. Proc Natl Acad Sci U S A. 2003;100(16):9302–7.

Bruder D, et al. Neuropilin-1: a surface marker of regulatory T cells. Eur J Immunol. 2004;34(3):623–30.

Brunkow ME, et al. Disruption of a new forkhead/winged-helix protein, scurfin, results in the fatal lymphoproliferative disorder of the scurfy mouse. Nat Genet. 2001;27(1):68–73.

Burchill MA, et al. IL-2 receptor beta-dependent STAT5 activation is required for the development of Foxp3+ regulatory T cells. J Immunol. 2007;178(1):280–90.

Carrier Y, et al. Th3 cells in peripheral tolerance. II. TGF-beta-transgenic Th3 cells rescue IL-2-deficient mice from autoimmunity. J Immunol. 2007a;178(1):172–8.

Carrier Y, et al. Th3 cells in peripheral tolerance. I. Induction of Foxp3-positive regulatory T cells by Th3 cells derived from TGF-beta T cell-transgenic mice. J Immunol. 2007b;178(1):179–85.

Charbonnier LM, et al. Immature dendritic cells suppress collagen-induced arthritis by in vivo expansion of CD49b+ regulatory T cells. J Immunol. 2006;177(6):3806–13.

Chen YH, et al. Regulatory T-cell clones induced by oral tolerance – suppression of autoimmune encephalomyelitis. Science. 1994;265(5176):1237–40.

Chen Y, et al. Lysine propionylation and butyrylation are novel post-translational modifications in histones. Mol Cell Proteomics. 2007;6(5):812–19.
Chen Z, Barbi J, Bu S, Yang HY, Li Z, Gao Y, et al. The ubiquitin ligase Stub1 negatively modulates regulatory T cell suppressive activity by promoting degradation of the transcription factor Foxp3. Immunity. 2013;39(2):272–85.
Dang EV, et al. Control of T(H)17/T(reg) balance by hypoxia-inducible factor 1. Cell. 2011;146(5):772–84.
DeBerardinis RJ, et al. The biology of cancer: metabolic reprogramming fuels cell growth and proliferation. Cell Metab. 2008;7(1):11–20.
De Rosa V, et al. A key role of leptin in the control of regulatory T cell proliferation. Immunity. 2007;26(2):241–55.
Feuerer M, et al. Enhanced thymic selection of FoxP3+ regulatory T cells in the NOD mouse model of autoimmune diabetes. Proc Natl Acad Sci U S A. 2007;104(46):18181–6.
Feuerer M, et al. Foxp3+ regulatory T cells: differentiation, specification, subphenotypes. Nat Immunol. 2009;10(7):689–95.
Floess S, et al. Epigenetic control of the foxp3 locus in regulatory T cells. PLoS Biol. 2007;5(2):169–78.
Fontenot JD, Gavin MA, Rudensky AY. Foxp3 programs the development and function of CD4+ CD25+ regulatory T cells. Nat Immunol. 2003;4(4):330–6.
Fontenot JD, et al. A function for interleukin 2 in Foxp3-expressing regulatory T cells. Nat Immunol. 2005;6(11):1142–51.
Fu W, et al. A multiply redundant genetic switch 'locks in' the transcriptional signature of regulatory T cells. Nat Immunol. 2012;13(10):972–80.
Fukaura H, et al. Induction of circulating myelin basic protein and proteolipid protein-specific transforming growth factor-beta 1-secreting Th3 T cells by oral administration of myelin in multiple sclerosis patients. J Clin Investig. 1996;98(1):70–7.
Gavin MA, et al. Single-cell analysis of normal and FOXP3-mutant human T cells: FOXP3 expression without regulatory T cell development (vol 103, pp. 6659, 2006). Proc Natl Acad Sci U S A. 2006;103(24):9373.
Gao Y, Lin F, Xu P, Nie J, Chen Z, Su J, et al. USP22 is a positive regulator of NFATc2 on promoting IL2 expression. FEBS Lett. 2014;588(6):878–83.
Gerriets VA, Rathmell JC. Metabolic pathways in T cell fate and function. Trends Immunol. 2012;33(4):168–73.
Gronborg M, et al. A mass spectrometry-based proteomic approach for identification of serine/threonine-phosphorylated proteins by enrichment with phospho-specific antibodies: identification of a novel protein, Frigg, as a protein kinase A substrate. Mol Cell Proteomics. 2002;1(7):517–27.
Groux H, et al. A CD4+ T-cell subset inhibits antigen-specific T-cell responses and prevents colitis. Nature. 1997;389(6652):737–42.
Haringer B, et al. Identification and characterization of IL-10/IFN-gamma-producing effector-like T cells with regulatory function in human blood. J Exp Med. 2009;206(5):1009–17.
Hinterberger M, Wirnsberger G, Klein L. B7/CD28 in central tolerance: costimulation promotes maturation of regulatory T cell precursors and prevents their clonal deletion. Front Immunol. 2011;2:30.
Hori S. Developmental plasticity of Foxp3(+) regulatory T cells. Curr Opin Immunol. 2010;22(5):575–82.
Hori S, Nomura T, Sakaguchi S. Control of regulatory T cell development by the transcription factor Foxp3. Science. 2003;299(5609):1057–61.
Horwitz DA, Zheng SG, Gray JD. Natural and TGF-beta-induced Foxp3(+)CD4(+) CD25(+) regulatory T cells are not mirror images of each other. Trends Immunol. 2008;29(9):429–35.
Jana S, et al. The role of NF-kappaB and Smad3 in TGF-beta-mediated Foxp3 expression. Eur J Immunol. 2009;39(9):2571–83.
Jeron A, et al. ChIP-on-chip analysis identifies IL-22 as direct target gene of ectopically expressed FOXP3 transcription factor in human T cells. BMC Genomics. 2012;13:705.

Jiang H, Chess L. The specific regulation of immune responses by CD8+ T cells restricted by the MHC class Ib molecule, Qa-1. Annu Rev Immunol. 2000;18:185–216.
Jin JO, Han X, Yu Q. Interleukin-6 induces the generation of IL-10-producing Tr1 cells and suppresses autoimmune tissue inflammation. J Autoimmun. 2013;40:28–44.
Jonuleit H, Schmitt E. The regulatory T cell family: distinct subsets and their interrelations. J Immunol. 2003;171(12):6323–7.
Jordan MS, et al. Thymic selection of CD4+ CD25+ regulatory T cells induced by an agonist self-peptide. Nat Immunol. 2001;2(4):301–6.
Kim HJ, Cantor H. Regulation of self-tolerance by Qa-1-restricted CD8(+) regulatory T cells. Semin Immunol. 2011;23(6):446–52.
Kretschmer K, et al. Inducing and expanding regulatory T cell populations by foreign antigen. Nat Immunol. 2005;6(12):1219–27.
Kuchen S, et al. Regulation of microRNA expression and abundance during lymphopoiesis. Immunity. 2010;32(6):828–39.
Levings MK, et al. Differentiation of Tr1 cells by immature dendritic cells requires IL-10 but not CD25+ CD4+ Tr cells. Blood. 2005;105(3):1162–9.
Li B, et al. FOXP3 interactions with histone acetyltransferase and class II histone deacetylases are required for repression. Proc Natl Acad Sci U S A. 2007;104(11):4571–6.
Li D, et al. T cell metabolism in autoimmune diseases. In: Chan J (ed) Autoimmune diseases – contributing factors, specific cases of autoimmune diseases, and stem cell and other therapies. InTech; 2012., 402pp. doi:10.5772/2896, ISBN 978-953-51-0693-7
Liang G, et al. Cooperativity between DNA methyltransferases in the maintenance methylation of repetitive elements. Mol Cell Biol. 2002;22(2):480–91.
Lu L, et al. All-trans retinoic acid promotes TGF-beta-induced Tregs via histone modification but not DNA demethylation on Foxp3 gene locus. PLoS ONE. 2011;6(9):e24590.
Magnani CF, et al. Killing of myeloid APCs via HLA class I, CD2 and CD226 defines a novel mechanism of suppression by human Tr1 cells. Eur J Immunol. 2011;41(6):1652–62.
Mahic M, et al. Generation of highly suppressive adaptive CD8(+)CD25(+)FOXP3(+) regulatory T cells by continuous antigen stimulation. Eur J Immunol. 2008;38(3):640–6.
Mardis ER. ChIP-seq: welcome to the new frontier. Nat Methods. 2007;4(8):613–14.
McKenna KC, Previte DM. Influence of CD8+ T regulatory cells on intraocular tumor development. Front Immunol. 2012;3:303.
Michalek RD, et al. Cutting edge: distinct glycolytic and lipid oxidative metabolic programs are essential for effector and regulatory CD4+ T cell subsets. J Immunol. 2011;186(6):3299–303.
Miyara M, Sakaguchi S. Natural regulatory T cells: mechanisms of suppression. Trends Mol Med. 2007;13(3):108–16.
Mosser DM, Zhang X. Interleukin-10: new perspectives on an old cytokine. Immunol Rev. 2008;226:205–18.
Mrowka R, Patzak A, Herzel H. Is there a bias in proteome research? Genome Res. 2001;11(12):1971–3.
Okamura T, et al. CD4+ CD25-LAG3+ regulatory T cells controlled by the transcription factor Egr-2. Proc Natl Acad Sci U S A. 2009;106(33):13974–9.
Oldenhove G, et al. Decrease of Foxp3(+) Treg cell number and acquisition of effector cell phenotype during lethal infection. Immunity. 2009;31(5):772–86.
Ono M, et al. Foxp3 controls regulatory T-cell function by interacting with AML1/Runx1. Nature. 2007;446(7136):685–9.
Ouyang W, et al. Novel Foxo1-dependent transcriptional programs control T(reg) cell function. Nature. 2012;491(7425):554–9.
Passerini L, et al. Functional type 1 regulatory T cells develop regardless of FOXP3 mutations in patients with IPEX syndrome. Eur J Immunol. 2011;41(4):1120–31.
Pot C, et al. Induction of regulatory Tr1 cells and inhibition of T(H)17 cells by IL-27. Semin Immunol. 2011;23(6):438–45.

Ramsdell F. Foxp3 and natural regulatory T cells: key to a cell lineage? Immunity. 2003;19(2): 165–8.
Ratushny V, Golemis E. Resolving the network of cell signaling pathways using the evolving yeast two-hybrid system. Biotechniques. 2008;44(5):655–62.
Rifa'i M, et al. Essential roles of CD8+ CD122+ regulatory T cells in the maintenance of T cell homeostasis. J Exp Med. 2004;200(9):1123–34.
Rouas R, et al. Human natural Treg microRNA signature: role of microRNA-31 and microRNA-21 in FOXP3 expression. Eur J Immunol. 2009;39(6):1608–18.
Rubtsov YP, et al. Stability of the regulatory T cell lineage in vivo. Science. 2010;329(5999):1667–71.
Rudra D, et al. Transcription factor Foxp3 and its protein partners form a complex regulatory network. Nat Immunol. 2012;13(10):1010–19.
Rust MJ, et al. Assembly of endocytic machinery around individual influenza viruses during viral entry. Nat Struct Mol Biol. 2004;11(6):567–73.
Sadlon TJ, et al. Genome-wide identification of human FOXP3 target genes in natural regulatory T cells. J Immunol. 2010;185(2):1071–81.
Sakaguchi S. Naturally arising CD4+ regulatory t cells for immunologic self-tolerance and negative control of immune responses. Annu Rev Immunol. 2004;22:531–62.
Samanta A, et al. TGF-beta and IL-6 signals modulate chromatin binding and promoter occupancy by acetylated FOXP3. Proc Natl Acad Sci U S A. 2008;105(37):14023–7.
Schmidt A, Oberle N, Krammer PH. Molecular mechanisms of treg-mediated T cell suppression. Front Immunol. 2012;3:51.
Shevach EM, et al. Role of TGF-Beta in the induction of Foxp3 expression and T regulatory cell function. J Clin Immunol. 2008;28(6):640–6.
Shi LZ, et al. HIF1alpha-dependent glycolytic pathway orchestrates a metabolic checkpoint for the differentiation of TH17 and Treg cells. J Exp Med. 2011;208(7):1367–76.
Thornton AM, et al. Expression of Helios, an Ikaros transcription factor family member, differentiates thymic-derived from peripherally induced Foxp3+ T regulatory cells. J Immunol. 2010;184(7):3433–41.
Tone Y, et al. Smad3 and NFAT cooperate to induce Foxp3 expression through its enhancer. Nat Immunol. 2008;9(2):194–202.
Tsun A, Chen Z, Li B. Romance of the three kingdoms: RORgammat allies with HIF1alpha against FoxP3 in regulating T cell metabolism and differentiation. Protein Cell. 2011;2(10):778–81.
Vang KB, et al. Cutting edge: CD28 and c-Rel-dependent pathways initiate regulatory T cell development. J Immunol. 2010;184(8):4074–7.
Vieira PL, et al. IL-10-secreting regulatory T cells do not express Foxp3 but have comparable regulatory function to naturally occurring CD4(+)CD25(+) regulatory T cells. J Immunol. 2004;172(10):5986–93.
Vlad G, et al. Immunoglobulin-like transcript 3-Fc suppresses T-cell responses to allogeneic human islet transplants in hu-NOD/SCID mice. Diabetes. 2008;57(7):1878–86.
Wang ET, et al. Alternative isoform regulation in human tissue transcriptomes. Nature. 2008;456(7221):470–6.
Wang Y, Su MA, Wan YY. An essential role of the transcription factor GATA-3 for the function of regulatory T cells. Immunity. 2011;35(3):337–48.
Weiss JM, et al. Neuropilin 1 is expressed on thymus-derived natural regulatory T cells, but not mucosa-generated induced Foxp3+ T reg cells. J Exp Med. 2012;209(10):1723–42, S1.
Wildin RS, et al. X-linked neonatal diabetes mellitus, enteropathy and endocrinopathy syndrome is the human equivalent of mouse scurfy. Nat Genet. 2001;27(1):18–20.
Wohlfert EA, et al. GATA3 controls Foxp3(+) regulatory T cell fate during inflammation in mice. J Clin Investig. 2011;121(11):4503–15.
Xie XS, Dunn RC. Probing single-molecule dynamics. Science. 1994;265(5170):361–4.
Xystrakis E, et al. Identification of a novel natural regulatory CD8 T-cell subset and analysis of its mechanism of regulation. Blood. 2004;104(10):3294–301.

Yadav M, et al. Neuropilin-1 distinguishes natural and inducible regulatory T cells among regulatory T cell subsets in vivo. J Exp Med. 2012;209(10):1713–22, S1–19.
Zabransky DJ, et al. Phenotypic and functional properties of Helios+ regulatory T cells. PLoS ONE. 2012;7(3):e34547.
Zhang Y, et al. Model-based analysis of ChIP-seq (MACS). Genome Biol. 2008;9(9):R137.
Zhang J, et al. Identification of the E3 deubiquitinase USP21 as a positive regulator of GATA3. J Biol Chem. 2013;288(13):9373–82.
Zheng SG, et al. IL-2 is essential for TGF-beta to convert naive CD4(+)CD25(−) cells to CD25(+)Foxp3(+) regulatory T cells and for expansion of these cells. J Immunol. 2007;178(4):2018–27.
Zheng Y, et al. Role of conserved non-coding DNA elements in the Foxp3 gene in regulatory T-cell fate. Nature. 2010;463(7282):808–12.
Zhou L, et al. TGF-beta-induced Foxp3 inhibits T(H)17 cell differentiation by antagonizing ROR gamma t function. Nature. 2008;453(7192):236-U14.
Zhou X, et al. Cutting edge: all-trans retinoic acid sustains the stability and function of natural regulatory T cells in an inflammatory milieu. J Immunol. 2010;185(5):2675–9.

Xuerui Luo is a Ph.D. student of Molecular Immunology Unit at Institut Pasteur of Shanghai (IPS), CAS. She was graduated from life Science College, Sichuan University in 2011. Recently she works in how PARP-1 regulates regulatory T cells through poly(ADP-ribosy)lation of FOXP3 under TGFβ/IL-6 signals.

Andy Tsun Group Leader/Senior Manager of Cell Line and New Drug Research and Development. Dr. Andy Tsun leads the teams for new drug development, lead drug selection, and cell line development. In collaboration with the early discovery company Adimab LLC, his team has developed functional assays and selected lead monoclonal antibody candidates for our PD-1 and PCSK-9 programs for subsequent generation of IgG-expressing CHO cell lines. Dr. Tsun also facilitates collaboration with academia in order to validate the functional activity of our drugs in vitro and in vivo.

Dr. Tsun has a master's degree from Imperial College London in Biochemistry, and obtained his Ph.D. from the University of Oxford in immune cell biology. Prior to joining Innovent Biologics Inc., he spent just under four years at Institut Pasteur of Shanghai where he was a Postdoctoral Fellow and then Associate Professor. Here, he developed T cell screening assays to test the effect of small molecule and biologic drug candidates on T cell function. He remains a Guest Associate Professor at Institut Pasteur of Shanghai.

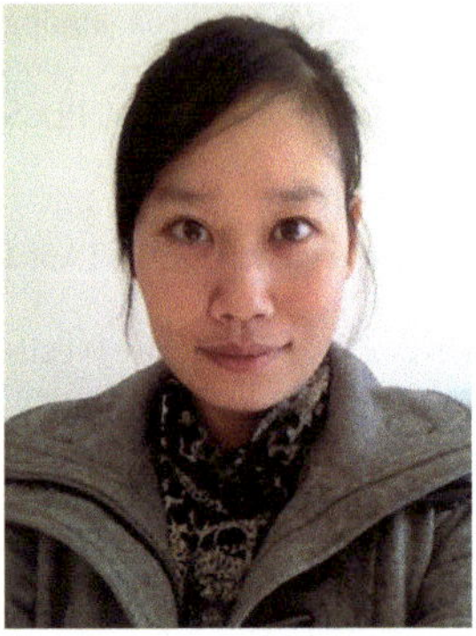

Jing Yang is a Ph.D. student of Molecular Immunology Unit at Institut Pasteur of Shanghai (IPS), CAS. She was graduated from life Science College, Shandong University in 2009. During her Ph.D. studies, she focus to understand the role of USP44 in regulatory T cells.

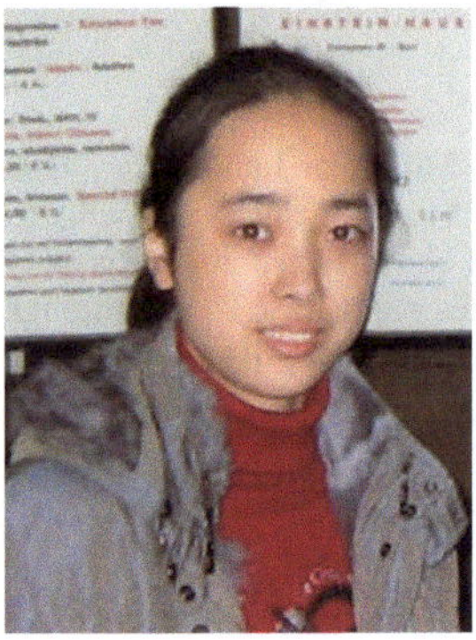

Zuojia Chen is a research fellow of Neurology Boston Department at Harvard Medical School. In 2014 she received her Ph.D. Degree in Institut Pasteur of Shanghai (IPS), CAS, China and she found STUB1 negatively modulates regulatory T cell suppressive activity by promoting FOXP3 degradation in Ph.D. period.

Yayi Gao is a research fellow of National Cancer Institute, USA. In 2014 she received her Ph.D. Degree in Institut Pasteur of Shanghai (IPS), CAS, China and she found DBC1 negatively regulated the function of regulatory T cells in Ph.D. period. She got her M.S. Degree in WUHAN Institute of Virology, CAS

Kongchen Li was graduated from Beijing Forestry University in 2010. Then she serves as a research assistant of Unit of Molecular Immunology in Institut Pasteur of Shanghai, Chinese Academy of Sciences. Her main research is focused on the molecular mechanism of regulatory T cells function.

Miranda Piccioni is an italian Biologist. In 2008 she received her Master of Science Degree in Biomedical and Molecular Sciences at University of Perugia, Italy, and she was awarded her PhD in 2012. During her doctoral studies, she analyzed the pathways of induction of immune tolerance by Cryptococcus neoformans capsular polysaccharides, examining how they affect host immunity under inflammatory and autoimmune conditions. In 2011 she joined the research group of Professor Bin Li, at Pasteur Institut of Shanghai, focusing on the mechanisms of differentiation and function of regulatory T cells. She was awarded the "Young International Scientists Fellowship" from the Chinese Academy of Sciences and published in peer-reviewed journals.

Bin Li is a Professor and Head of Molecular Immunology Unit at Institut Pasteur of Shanghai (IPS), CAS. In 2001, he received his Doctor of Philosophy (Ph.D.) degree in Peking University, College of Life Sciences, China. In 2009 he joined the Pasteur Institut of Shanghai, CAS. His main research is focused to understand and therapeutically modulate the stability of regulatory T cells to treat inflammatory diseases.

His group interests in elucidating the cellular and molecular mechanisms regulating the immune response in health and disease including infectious diseases, autoimmune diseases and cancer. Currently, they are particularly interested in understanding the dynamic role of one subpopulation of T cells, namely CD4+CD25+FOXP3+ regulatory T cells (FOXP3+Tregs), in major human infectious diseases.

Chapter 7
γδT Cells and αβT Cells

Lingyan Wang, Fuping Wang, Jieyi Shi, and Xiangdong Wang

Abstract T cells were divided into two groups by T cell receptors (TCRs), αβT cells and γδT cells. αβT cells, with T cells were divided into two groups by T cell receptors (TCRs), αβT cells and γδT cells. αβT cells, with TCR2, mainly mediate cell immunity and immune-regulation while γδT cells, with TCR1, play important functions in wound healing, removing distressed or transformed epithelial cells and subduing excessive inflammation besides maintaining immune homeostasis in the local microenvironment. Studies have revealed that αβT cells and γδT cells play different roles in autoimmune diseases, tumors and vascular diseases etc. Interestingly, αβT cells and γδT cells have a close interactions with each other in maintaining human body balance to defence diseases. In this review, we concluded the introduction of αβT cells and γδT cells with their specific biomarkers, signalling pathways, protein and geneomix and functions in disease progression and prevention.

Keywords αβT cells • γδT cells • Autoimmune diseases • Cancer • Omix

7.1 Introduction

T cells were divided into two groups by T cell receptors (TCRs), αβT cells and γδT cells. αβT cells, express TCR2, consist of 65–75 % of PBMC in peripheral blood while γδT cells account for less than 10 %. They express different surface

L. Wang
Fudan University Center for Clinical Bioinformatics, Zhongshan Hospital Biomedical Research Center, Zhongshan Hospital, Fudan University, Shanghai 200032, China

F. Wang
Department of Gastroenterology, Zhongshan Hospital, Fudan University, Shanghai 200032, China

J. Shi
Department of Liver Surgery, Zhongshan Hospital, Fudan University, Shanghai 200032, China

X. Wang, M.D., Ph.D. (✉)
Zhongshan Hospital, Fudan University, China

Shanghai Institute of Clinical Bioinformatics, China
e-mail: xiangdong.wang@clintransmed.org

X. Wang (ed.), *Single Cell Sequencing and Systems Immunology*,
Translational Bioinformatics 5, DOI 10.1007/978-94-017-9753-5_7

markers of CD4 and CD8, e.g. 60 % αβT cells are CD4 positive, 30 % CD8 positive, and both positive less than 1 % in αβ T cells. The majority of γδT cells are double negative (Kalyan and Kabelitz 2013). Due to the role as co-receptors for MHC class I or MHC class II molecules, CD8 and CD4 expression dictates the MHC restriction pattern of the αβT cells, while not in γδT cells in the antigen recognition behavior (Kabelitz and He 2012; Hayday 2000). TCR2 consists of α, β peptide in αβT cells, diversified owing to the large number of Vα and Vβ gene rearrangement. In contrast, TCR1, composed of γ, δ peptide, only has a few Vγ and Vδ germline genes in constructing functional γδ TCRs. Different constructions of peptides as a primary effector of the adaptive immune response contribute to diverse functions of αβT cells and γδT cells, mainly mediate cell immunity and immune-regulation. A large population of intraepithelial cytolytic αβT cells in human small intestine play a main role in the maintenance of the epithelial integrity by rapidly eliminating infected and damaged cells (Jabri and Ebert 2007). $CD3^+CD4^-CD8^-$ αβT cells have a capability to inhibit the function of $CD8^+$ T cells carrying the same T cell receptor specificity and prevent the rejection of skin allografts (Zhang et al. 2001). This indicated that despite the traditional αβT cells, other unique T cells that express α and β peptide also have the function of traditional αβT cells. Opposing αβT cells are particularly well equipped to respond to the "missing self" and the "dangerous non-self", γδT cells, enriched in skin, the digestive tract and reproductive organ mucosa, have a highly empathetic nature to deal with the "safe non-self" and "distressed self" situation, which hangs the balance of our body (Kalyan and Kabelitz 2013; Karre 1997). In addition to their role of γδT cells in maintaining immune homeostasis in the local microenvironment, tissue-associated γδT cells, in both mice and men, play an important function in wound healing, removing distressed or transformed epithelial cells and subduing excessive inflammation (Bonneville 2012). Fan et al. found that Human Vδ1 T cells have been observed to be recruited to the maternal/fetal interface during pregnancy where their activation is believed to play a role in the induction of tolerance to paternal antigens (Barakonyi et al. 2002; Fan et al. 2011). Although αβ/γδT cells have their own specific functions in the local area, they definitely closely contact with each other in human immunity system to balance the body health. Studies on γδT cells in the lung have shown that γδT cells influence B and T cell responses. For instance, γδT cells modulate $CD4^+$αβT cells in tuberculosis, and in allergic inflammation they support IgE production (Lahn 2000). Bosco N group's result also reported that TCR-β chains from γδ T cells can efficiently take part in αβT-cell development (Bosco et al. 2008). These interesting phenomena displays the complex system between the function of αβ/γδT cells, which needs further investigated.

7.2 Cell Receptors and Functions

Antigen recognition by T lymphocytes is mediated by cell-surface glycol-proteins known as T-cell antigen receptors (TCRs). These composed of α and β, or γ and δ, polypeptide chains with variable (V) and constant (C) regions (Li et al. 1998).

Since the first crystal structure determinations of αβ T cell receptors (TCRs) bound to class I MHC-peptide (pMHC) antigens in 1996, a sizable database of 24 class I and class II TCR/pMHC complexes has been accumulated that now defines a substantial degree of structural variability in TCR/pMHC recognition (Rudolph et al. 2006). And CD8 is proposed to act as a time and distance lock, enabling the shorter TCR molecule to bridge the pMHC and have time to finely read the peptide (Puech et al. 2011). Specific recognition of peptides is required to trigger developmentally appropriate responses at the distinct stages of T-cell maturation. For example, developing T cells in the thymus differ from peripheral T cells in that they possess substantially fewer TCRs on the surface, express both the CD4 and CD8 co-receptors, and make use of thymic epithelial cells to present self-antigens for at least the positive selection events. Furthermore, peripheral T cells interact with several types of antigen-presenting cells (APCs) and target cells expressing antigens as they mature into specific effector and memory cells. Thus, the same clonal TCR can lead to a range of T-cell responses and different cell fates depending on the context of interaction with the antigen (Zhu et al. 2013). Opposite to the αβT cells, γδ TCRs appear to recognize proteins directly, without antigen processing, and to recognize MHC molecules independently of the bound peptide. Moreover, small phosphate-containing non-peptide compounds have also been identified as ligands for certain γδT cells (Li et al. 1998). In some other researches, human γδT cells also express activating receptors, such as NKG2D that recognizes stress-inducible MHC class I-related MICA/MICB molecules and the UL16-binding proteins that are up-regulated on malignant or stressed cells and induce cytolysis (Wrobel et al. 2007). These two different talents of receptor and ligand binding in αβ and γδT cells disturb the various functions on human health, which maintains it in a balance situation.

7.3 Cell Signaling

7.3.1 ERK Pathway

TCR signaling leads to ERK/MAP kinase (MAPK) activation and is required for normal thymocyte development (Lee et al. 2010; Fischer et al. 2005). Previous studies established that Erk1/2 MAPKs are critical for positive selection of αβT cells. Additional evidence suggests that increased Erk1/2 activity promotes γδT cell generation (Chen et al. 2012; Haks et al. 2005; Hayes et al. 2005). When ERK/MAPK activation, it not only phosphorylates regulatory targets in the cytosol, but also activates the key genes in nucleus, such as Elk-1, c-Fos, or c-Myc (Roux and Blenis 2004). Frascoli M group found ATP release and P2X7 signaling upon γδ TCR expression in immature thymocytes constitutes an important costimulus in T cell lineage choice through the ERK-Egr-Id3 signaling pathway and contributes to shaping the peripheral γδT cell compartment (Frascoli et al. 2012). Accordingly, Haks et al. revealed that Egr induction correlates with signal strength during αβ/γδ lineage development and elevating Egr levels through ectopic expression augmented development of γδ lineage cells while causing a commensurate decrease in αβ lineage cells (Lee et al. 2010; Haks et al. 2005; Lauritsen et al. 2009). Another experiment also demonstrated that when in the absence of Id3, numbers of αβT cells were generated even when the γδ TCR ligand was expressed (Lauritsen et al. 2009). This suggests that the ERK-Egr-Id3 signaling pathway plays a vital role in the basis development of αβT cells and γδT cells (Kreslavsky et al. 2010).

7.3.2 Notch Pathway

Notch signals, involved in various organism courses such as stem cell differentiation (Martini et al. 2013), cancer development (Jonusiene et al. 2013), are initially thought to determine the αβ versus γδ lineage fate (Tanigaki et al. 2004) (Washburn et al. 1997). Researches have found that γδ lineage thymocytes become Notch-independent only on expression of γδ TCR complex, while αβ lineage precursors are dependent on Notch signaling throughout the entirety of their differentiation (Lee et al. 2010; Ciofani et al. 2006). That may indicate that αβ lineage fate is more reliable on the Notch signaling. Furthermore, experiments in recent years suggest that the dependence of Notch signaling in αβ and γδ T cells development somehow related to the E protein suppression (Nie et al. 2003; Ikawa et al. 2006).

7.3.3 CXCR4 Pathway

The β-selection checkpoint is important in αβT cells development. Once unreached the threshold, αβT cells can not develop in a feeder-free system. Researches in recent years have been indicated that both pre-TCR and Notch signaling are not sufficient for the αβ lineage development, which should combine the CXCR4 signal, required for transition through the β-selection checkpoint (Dallas et al. 2005; Varnum-Finney et al. 2003; Trampont et al. 2010; Janas et al. 2010). A complementary approach revealed that downstream of CXCR4, the phosphoinositide 3-kinase (PI3K) subunit isoforms p110γ and p110δ were both required for optimal αβT cells progression (Janas et al. 2010). This indicates that the CXCR4-PI3K-Akt pathway do the key role in αβT cell progression.

7.3.4 Other Pathways

Despite the pre-TCR, Notch and CXCR4 pathway's vital role in the balance of αβ/γδT cells development, some other pathways also do a complementary part in this action. Zocchi et al. revealed that the TGF- β mediated NKG2D pathway effects on the $CD8^+$αβT and γδT cells' pathological feedback (Zocchi et al. 2012). The Staal FJ goup showed convincing evidence that canonical Wnt signaling, which requires the stabilization of β-catenin, was crucial for normal T-cell development

(Staal et al. 2001). Also, the IL-17A mediated pathway and the TNF-α signaling somehow do great effect on αβ/γδ lineage establishment (Chatterjee et al. 2012; Kelsen et al. 2011).

7.4 Cell Biomarkers

Analysis of the molecular control of αβ/γδ T lineage commitment continues to be hampered by the lack of definitive lineage markers of the early stages of commitment in thymocytes. The TCR complexes (i.e., pre-TCR expression for αβ lineage and γδTCR for γδ lineage) have also been employed as lineage markers, but the TCR isotype alone is no longer sufficient to assign lineage fate for development of different T subsets. For example, both the pre-TCR and γδTCR are able to support αβ lineage commitment and development to the double positive (DP) stage (Lee et al. 2010). Accordingly, CD4 and CD8 expression must also be taken into consideration in assigning lineage, such that γδTCR expressing cells that remain double negative (DN) are assigned to the γδ lineage, while those developing to the DP stage in response to TCR signals from any receptor isotype are assigned to the αβ lineage. Recently, many other cell biomarkers to determine the αβ/γδT cells have been established. Kreslavsky T's group has utilized downregulation of CD24 (HSA) among DNs as an additional marker of γδ commitment (Kreslavsky et al. 2008). More efforts identified the transcription factor Sox13 being highly enriched in γδ T cells (Spicuglia et al. 2007) while Foxp3 is a unique transcription factor of Treg, a subset of αβT cells (Chen et al. 2013). Though efforts to gain insight into the molecular control of αβ/γδ lineage commitment would benefit from definitive molecular markers distinguishing DN thymocytes committing to the αβ lineage from those committing to the γδ lineage, such markers remain elusive and need further investigated.

7.5 Protein Omix

We found that the proteins related to γδT cells progression are about 15 reported while αβT cells display around 46 proteins. For example, Hoelbl A group highlighted that the STAT5 transcription factor have been implicated in lymphoid development and transformation (Hoelbl et al. 2006). Another group also stated that MEK1/2 induces TCRγ germline transcription by phosphorylating STAT5 through IL-7R-FFFF (Maki and Ikuta 2008). They all indicated that STAT5 have an impact on the development of γδT cells. Other evidence pointed to a role for the CD277/butyrophilin-3 (BTN3A) molecules in the activation of γδT cells population, which bind to an agonist antibody (20.1) that mimics phosphoisoprenoid-mediated Vγ9Vδ2, a large amount γδT cells in human, of activation and antagonist antibody (103.2) which inhibits this reactivity (Harly et al. 2012; Palakodeti et al. 2012).

Kadlecek et al. demonstrated that the Syk/ZAP-70 family of protein tyrosine kinases is indispensable for normal lymphoid development. Syk is necessary for the development of epithelial γδT cells, whereas ZAP-70 is essential for the normal development of αβT cells (Kadlecek et al. 1998). In order to conclude these proteins, we use the String and Network Ontology Analysis software to see the connections between them. Results showed that the proteins are mainly related to immune response-activating cell surface receptor signaling pathways, positive regulation of immune system process and leukocyte activation. The protein connections of αβ/γδT cells were shown in Figs. 7.1 and 7.2, indicating SYK-ZAP70-BCL2-SATB1/PTPRC-FOXP3-GATA3-FUT7 and other pathways maybe influence the αβT cells development while KLRK1-PTPRC-LCK-STAT5B and rest pathways do some meanings on γδT cells progression.

7.6 Genomix

Gene expression profiling is a powerful tool for establishing a molecular map for lymphocyte. Kana Miyazaki group recently used the microarray to detect the αβ/γδ T-cell lymphoma(TCL) differences in gene expression (Miyazaki et al. 2009). They found that in γδTCL, genes of natural killer (NK) cell–associated molecules, such as killer cell immunoglobulin (Ig)–like receptor (KIR) genes (KIR3DL1, KIR2DL4, and KIR2DL2), and killer cell lectin-like receptors (KLRC4, KLRD1, and KLRC2) were found to be over-expressed (Table 7.1). This result is closely accordant with the previous studies (Wrobel et al. 2007). To further search for functionally important genes over-expressed in αβ/γδ TCL, they performed GO and pathway analysis. The result displayed that the enriched GO categories in γδTCL were cellular defense response, signal transduction activity, receptor activity, transmembrane receptor activity, and IgG binding while in αβTCL were organismal physiologic process, regulation of organismal physiologic process, circulation, regulation of neurophysiologic process, regulation of transmission of nerve impulse, chemotaxis, receptor binding and so on (Table 7.1). They all suggested that the different states of αβ/γδ in various sides, which provide us an integrated view on T cell acting on human body.

7.7 Role in Diseases

T cells, as the major roles in the immune system, play important roles in many diseases such as infections, tumor, autoimmune diseases and so on. Different subsets of T cells all have their direct or indirect relations with the pathogenesis and therapies of those diseases. Involving virus infection, CD4 T-lymphocytes are the primary target cells for human immunodeficiency virus (HIV). The progressive depletion of these cells eventually results in weakening of the host's immune ability

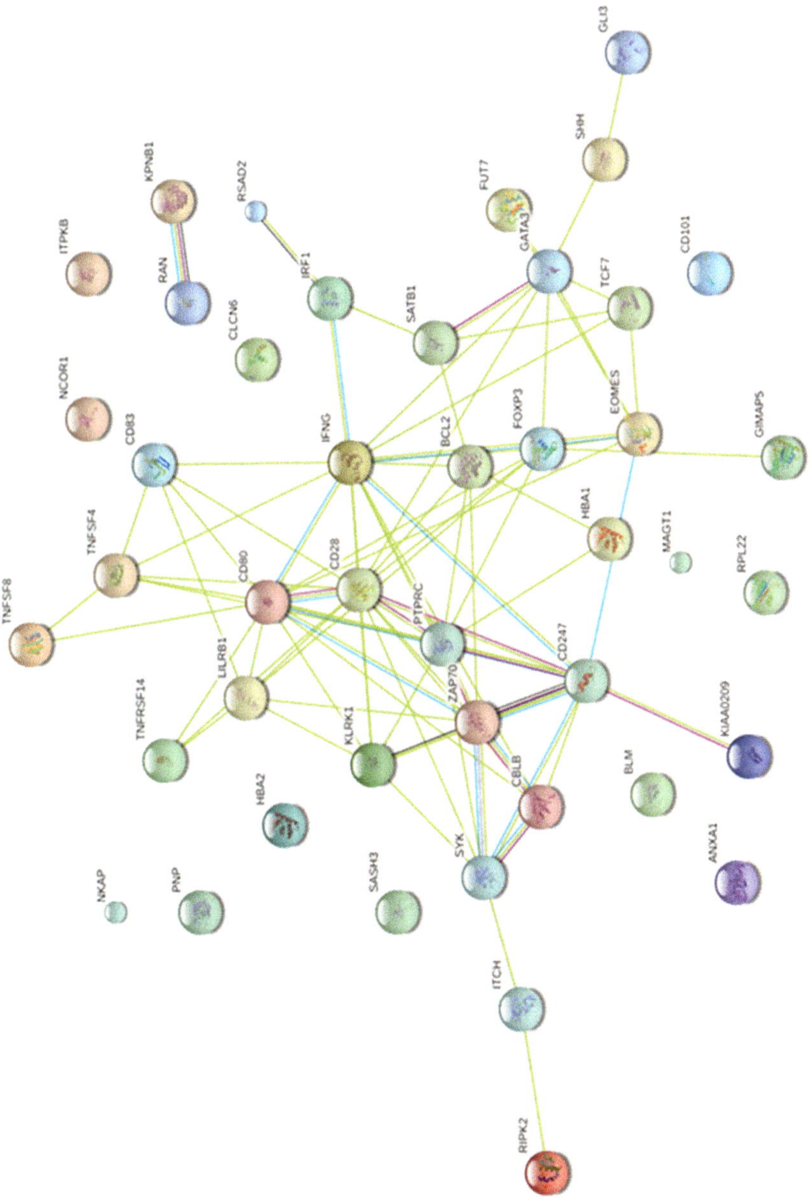

Fig. 7.1 The protein connections of αβT cells

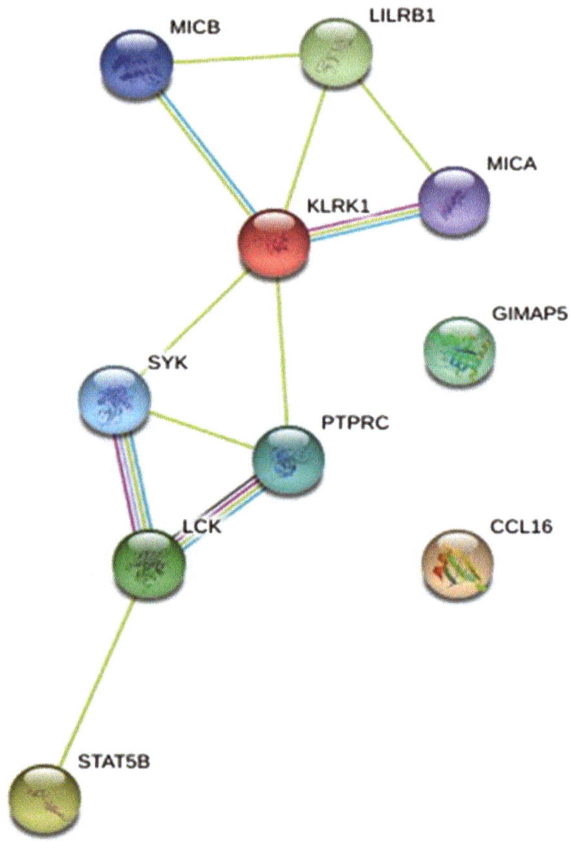

Fig. 7.2 The protein connections of γδT cells

to fight against any pathogen, thus rendering the host susceptible to infections and leading ultimately to death of patients in the terminal stage of acquired immune deficiency syndrome (AIDS) (Pattanapanyasat 2012).

Actually with the new findings of specific subsets of T cells, Tregs and Th17 cells demonstrated important functions regardless of their low frequency. Named after their ability of producing IL-17 or IL-10, TGF-beta, Th17 cells participated in many kinds of diseases. Recent studies show that Th17 cells and IL-17 play a pathogenic role in the development of various autoimmune diseases, including EAE, inflammatory bowel disease (IBD), inflammatory skin diseases, and collagen-induced arthritis (CIA), as well as graft-versus-host disease (Sutton et al. 2006; Plater-Zyberk et al. 2009). γδT cells, together with natural killer (NK) cells, NKT cells, and several populations of innate lymphoid cells (ILCs), belong to a family of IL-17-secreting lymphocytes that fits more closely with the innate rather than the adaptive immune system. The Vγ4 subset of γδ T cells has been shown to be a major source of IL-17 in Con-A-induced fulminant hepatitis (Zhao et al. 2011), while the Vγ4 subset of γδ T cells has been shown to control RSV and West Nile virus infection (Welte et al. 2011; Dodd et al. 2009).

Table 7.1 GO category analysis in the TCR signature gene set

Analytical tool	Gene No.	Gene	P
Go category			
γδTCL			
Cellular defense response	5	KIR2DL4, NCR1, C3AR1, KLRC2, KLRC4	$1.17*10^{-4}$
Signal transduction activity	29	KLRD1, ANXA9, FNDC3B, GPR37, KIR2DL4, MARCO, EDG7, NCR1, MS4A5,	$5.09*10^{-6}$
		HPGD, FCGR3A, LGR4, FCRLB, C3AR1, CXCL12, RTN4R, PAQR9,	
		KLRC2, GPR153, FCGR3B, KIR2DL2, EDG8, ADRB1, CD36, FZD5,	
		SCARF2, KIR3DL1, EPHA6, KLRC4	
Receptor activity	28	GPR37, KIR2DL4, ANXA9, KLRD1, FNDC3B, EDG7, HPGD, MARCO, LGR4,	$7.10*10^{-8}$
		MS4A5, FCGR3A, NCR1, C3AR1, FCRLB, KLRC2, PAQR9, RTN4R,	
		GPR153, EDG8, FCGR3B, KIR2DL2, ADRB1, FZD5, CD36, SCARF2,	
		KIR3DL1, EPHA6, KLRC4	
Transmembrane receptor activity	16	ANXA9, GPR37, KIR2DL4, KLRD1, LGR4, HPGD, MARCO, EDG7, C3AR1,	$4.46*10^{-4}$
		KLRC2, EDG8, GPR153, ADRB1, FZD5, EPHA6, KIR3DL1	
IGG binding	2	FCGR3A, FCGR3B	$3.07*10^{-4}$
αβ TCL			
Organismal physiologic process	14	UBD, CXCL13, COL4A4, CCL18, KCNE2, CCL17, C3, TMEM142A, DLL4,	$3.24*10^{-4}$
		APOE, COL4A3, TNFRSF25, MMP9, CCL19	
Regulation of organismal physiologic process	4	COL4A4, KCNE2, C3, APOE	$5.87*10^{-4}$
Circulation	4	KCNE2, DLL4, COL4A3, APOE	$3.14*10^{-4}$
Regulation of neurophysiologic process	2	COL4A4, APOE	$7.55*10^{-4}$
Regulation of transmission of nerve impulse	2	COL4A4, APOE	$7.55*10^{-4}$
Regulation of synapse structure and function	2	COL4A4, APOE	$8.69*10^{-4}$
Regulation of synaptic transmission	2	COL4A4, APOE	$7.55*10^{-4}$
Inflammatory response	5	CXCL13, CCL18, CCL17, C3, CCL19	$6.05*10^{-4}$
Behavior	5	CCL18, CXCL13, CCL17, APOE, CCL19	$2.47*10^{-4}$
Locomotory behavior	4	CXCL13, CCL18, CCL17, CCL19	$4.92*10^{-4}$
Taxis	4	CXCL13, CCL18, CCL17, CCL19	$4.22*10^{-4}$

(continued)

Table 7.1 (continued)

Analytical tool	Gene No.	Gene	P
Chemotaxis	4	CXCL13, CCL18, CCL17, CCL19	$4.22*10^{-4}$
Receptor binding	9	ADAMDEC1, CCL18, CXCL13, CCL17, C3, DLL4, APOE, COL4A3, CCL19	$2.76*10^{-5}$
G-protein–coupled receptor binding	4	CCL18, CXCL13, CCL17, CCL19	$1.57*10^{-5}$
Chemokine receptor binding	4	CCL18, CXCL13, CCL17, CCL19	$6.97*10^{-6}$
Chemokine activity	4	CCL18, CXCL13, CCL17, CCL19	$6.37*10^{-6}$
Extracellular region	11	CCL18, COL4A4, CXCL13, CCL17, C3, WNT5B, MMP9, COL4A3, APOE, SPOCK2, CCL19	$7.97*10^{-5}$
Extracellular region part	9	CCL18, COL4A4, CXCL13, CCL17, SPOCK2, APOE, COL4A3, MMP9, CCL19	$7.33*10^{-5}$
Sheet-forming collagen	2	COL4A4, COL4A3	$1.70*10^{-4}$
Collagen type IV	2	COL4A4, COL4A3	$1.21*10^{-4}$

Table from Miyazaki et al. (2009)

Yδ T cells also play a pathological role in certain autoimmune diseases, where they provide an early source of IL-17 and IL-21, which initiate responses mediated by conventional IL-17-secreting CD4$^+$ T cells (Th17 cells). The discovery of these innate sources of IL-17 has led to a reexamination of the roles played by effector and pathogenic cells in diseases where IL-17 is implicated, such as bacterial and fungal infection and cancer, as well as in gut homeostasis. In addition, these innate IL-17 producers have been shown to participate in the initiation of autoimmune diseases including experimental autoimmune encephalomyelitis (EAE), arthritis, and colitis (Sutton et al. 2009; Geremia et al. 2011; Roark et al. 2007). Whether the IL-17 producing cells are benefit or inhibitory for the disease progression, there is no absolute conclusions. T cells have a pathogenic role in autoimmunity, it has also been shown that intraepithelial γδ T cells play a protective role in dextran sodium sulfate (DSS)-induced colitis by preserving the integrity of the intestinal epithelium (Chen et al. 2002) although the mechanistic explanations for these different roles are currently unknown. What is subject to believed is that L-17 and IL-21 production by γδ T cells may help to initiate or augment IL-17 production by αβ T-cell activation, thus γδ T cells may act to prime Th17-cell responses (Sutton et al. 2009).

For cancer, a relationship between higher levels of circulating responsive γδT cells and a more favorable outcome has been observed, γδ T-cell IL-17 production, induced by dying tumor cells during chemotherapy (Ma et al. 2011), was also found to be essential for the control of tumor growth via chemoattraction of CD8$^+$ T cells and subsequent CD8$^+$T-cell IFN-γ production (Ma et al. 2011). The ability of γδ T cells to act in an APC-like manner has been exploited in their use as immunotherapeutics for cancer. The aim of cancer immunotherapy is to overcome immunosuppression at the site of the tumor by skewing the cytokine repertoire in favor of proinflammatory responses. Ex vivo activated γδ T cells have been shown to control tumor growth (Zhou et al. 2012), while clinical trials with activated γδ T cells have shown promising results for the treatment of solid tumors (Bennouna et al. 2010), lymphoma (Zhou et al. 2012), renal carcinoma (Kobayashi et al. 2007), and lung cancer (Noguchi et al. 2011).

Besides the above familiar T cell-related diseases, vascular or coronary diseases have been focused on the immune mechanisms. Previous opinions on vascular disease is the imbalanced metabolism of fat in the vessel wall, however a new "protein homeostasis system" of the host has been proposed and hyperactive immune cells, especially T cells, with excessive cytokines may be responsible for tissue injury as well as for tissue reconstruction according to this hypothesis. Kawasaki disease (KD) is a self-limited systemic inflammatory illness, and coronary artery lesions (CALs) are a major complication determining the prognosis of the disease. Innate and adaptive immune cells control pathogenic proteins that are toxic to host cells at a molecular level. After an infection of unknown KD pathogen(s), the pathogenic proteins initially attracted non-specific T cells and non-specific antibodies and then hyper-activated immune cells produce various cytokines, leading to a cytokine imbalance associated with further endothelial cell injury. After the emergence of specific T cells and specific antibodies against the pathogenic proteins, tissue injury ceases and a repair reaction begins with the immune cells (Lee et al. 2012). Besides, Tregs also play important roles in coronaviral infections. What is different is that arteriviruses have exploited these T cell subsets as a means to dampen the immune response allowing for viral persistence, while induction of Tregs in coronaviral infections protects against the more severe forms of the disease attributable to the host response (Cecere et al. 2012).

Above all indicate that γδ T cells, activated by different stimulation, can produce IL-17 to induce the adaptive immune responses mediated by different subsets of αβ T cells. Not exceptional, γδ T cells are an important ingraft source of IFN-γ and IL-17a and inhibit the production of inflammation cytokine by anti-p40 antibody. The graft survival time in the γδ T cell receptor (−/−) and IL-17(−/−) mice was prolonged significantly. In the chronic allograft rejection model, treatment with anti-p40 antibody prolongs graft survival possibly by reducing the amount of reactive inflammatory cells, especially γδ T cells (Wang et al. 2012). And in the future investigation on mechanisms of diseases, T cells, both αβ T cells and γδ T cells should be studied as a whole to explain mysteries behind diseases.

7.8 Conclusion

However, it is interesting to note that more and more findings of γδT cells and αβT cells have been revealed, helping us to understand these special functions on human body. As human beings is a integrated living beings, the interactions between γδT cells and αβT cells and their inter-works with other cells and organs should be considered together for further researches and produce more benefits to human.

References

Barakonyi A, et al. Recognition of nonclassical HLA class I antigens by gamma delta T cells during pregnancy. J Immunol. 2002;168(6):2683–8.
Bennouna J, et al. Phase I study of bromohydrin pyrophosphate (BrHPP, IPH 1101), a Vgamma9Vdelta2 T lymphocyte agonist in patients with solid tumors. Cancer Immunol Immunother. 2010;59(10):1521–30.
Bonneville M. Semaphorins: new cues for skin healing by gammadelta T cells. Immunity. 2012;37(2):194–6.
Bosco N, et al. TCR-beta chains derived from peripheral gammadelta T cells can take part in alphabeta T-cell development. Eur J Immunol. 2008;38(12):3520–9.
Cecere TE, Todd SM, Leroith T. Regulatory T cells in arterivirus and coronavirus infections: do they protect against disease or enhance it? Viruses. 2012;4(5):833–46.
Chatterjee M, et al. CD3-T cell receptor co-stimulation through SLAMF3 and SLAMF6 receptors enhances RORgammat recruitment to the IL17A promoter in human T lymphocytes. J Biol Chem. 2012;287(45):38168–77.
Chen Y, et al. Protection of the intestinal mucosa by intraepithelial gamma delta T cells. Proc Natl Acad Sci U S A. 2002;99(22):14338–43.
Chen Y, et al. Differential requirement of RasGRP1 for gammadelta T cell development and activation. J Immunol. 2012;189(1):61–71.
Chen X, et al. TNFR2 is critical for the stabilization of the CD4+Foxp3+ regulatory T cell phenotype in the inflammatory environment. J Immunol. 2013;190(3):1076–84.
Ciofani M, et al. Stage-specific and differential notch dependency at the alphabeta and gammadelta T lineage bifurcation. Immunity. 2006;25(1):105–16.
Dallas MH, et al. Density of the Notch ligand Delta1 determines generation of B and T cell precursors from hematopoietic stem cells. J Exp Med. 2005;201(9):1361–6.
Dodd J, et al. Pulmonary V gamma 4+ gamma delta T cells have proinflammatory and antiviral effects in viral lung disease. J Immunol. 2009;182(2):1174–81.
Fan DX, et al. The decidual gamma-delta T cells up-regulate the biological functions of trophoblasts via IL-10 secretion in early human pregnancy. Clin Immunol. 2011;141(3):284–92.
Fischer AM, et al. The role of erk1 and erk2 in multiple stages of T cell development. Immunity. 2005;23(4):431–43.
Frascoli M, et al. Purinergic P2X7 receptor drives T cell lineage choice and shapes peripheral gammadelta cells. J Immunol. 2012;189(1):174–80.
Geremia A, et al. IL-23-responsive innate lymphoid cells are increased in inflammatory bowel disease. J Exp Med. 2011;208(6):1127–33.
Haks MC, et al. Attenuation of gammadeltaTCR signaling efficiently diverts thymocytes to the alphabeta lineage. Immunity. 2005;22(5):595–606.
Harly C, et al. Key implication of CD277/butyrophilin-3 (BTN3A) in cellular stress sensing by a major human gammadelta T-cell subset. Blood. 2012;120(11):2269–79.
Hayday AC. [gamma][delta] cells: a right time and a right place for a conserved third way of protection. Annu Rev Immunol. 2000;18:975–1026.
Hayes SM, Li L, Love PE. TCR signal strength influences alphabeta/gammadelta lineage fate. Immunity. 2005;22(5):583–93.
Hoelbl A, et al. Clarifying the role of Stat5 in lymphoid development and Abelson-induced transformation. Blood. 2006;107(12):4898–906.
Ikawa T, et al. E proteins and Notch signaling cooperate to promote T cell lineage specification and commitment. J Exp Med. 2006;203(5):1329–42.
Jabri B, Ebert E. Human CD8+ intraepithelial lymphocytes: a unique model to study the regulation of effector cytotoxic T lymphocytes in tissue. Immunol Rev. 2007;215:202–14.
Janas ML, et al. Thymic development beyond beta-selection requires phosphatidylinositol 3-kinase activation by CXCR4. J Exp Med. 2010;207(1):247–61.

Jonusiene V, et al. Down-regulated expression of Notch signaling molecules in human endometrial cancer. Med Oncol. 2013;30(1):438.
Kabelitz D, He W. The multifunctionality of human Vgamma9Vdelta2 gammadelta T cells: clonal plasticity or distinct subsets? Scand J Immunol. 2012;76(3):213–22.
Kadlecek TA, et al. Differential requirements for ZAP-70 in TCR signaling and T cell development. J Immunol. 1998;161(9):4688–94.
Kalyan S, Kabelitz D. Defining the nature of human gammadelta T cells: a biographical sketch of the highly empathetic. Cell Mol Immunol. 2013;10(1):21–9.
Karre K. How to recognize a foreign submarine. Immunol Rev. 1997;155:5–9.
Kelsen J, et al. Infliximab induces clonal expansion of gammadelta-T cells in Crohn's disease: a predictor of lymphoma risk? PLoS ONE. 2011;6(3):e17890.
Kobayashi H, et al. Safety profile and anti-tumor effects of adoptive immunotherapy using gamma-delta T cells against advanced renal cell carcinoma: a pilot study. Cancer Immunol Immunother. 2007;56(4):469–76.
Kreslavsky T, et al. T cell receptor-instructed alphabeta versus gammadelta lineage commitment revealed by single-cell analysis. J Exp Med. 2008;205(5):1173–86.
Kreslavsky T, et al. alphabeta versus gammadelta fate choice: counting the T-cell lineages at the branch point. Immunol Rev. 2010;238(1):169–81.
Lahn M. The role of gammadelta T cells in the airways. J Mol Med (Berl). 2000;78(8):409–25.
Lauritsen JP, et al. Marked induction of the helix-loop-helix protein Id3 promotes the gammadelta T cell fate and renders their functional maturation Notch independent. Immunity. 2009;31(4):565–75.
Lee SY, et al. Towards a molecular understanding of the differential signals regulating alphabeta/gammadelta T lineage choice. Semin Immunol. 2010;22(4):237–46.
Lee KY, Rhim JW, Kang JH. Kawasaki disease: laboratory findings and an immunopathogenesis on the premise of a "protein homeostasis system". Yonsei Med J. 2012;53(2):262–75.
Li H, et al. Structure of the Vdelta domain of a human gammadelta T-cell antigen receptor. Nature. 1998;391(6666):502–6.
Ma Y, et al. Contribution of IL-17-producing gamma delta T cells to the efficacy of anticancer chemotherapy. J Exp Med. 2011;208(3):491–503.
Maki K, Ikuta K. MEK1/2 induces STAT5-mediated germline transcription of the TCRgamma locus in response to IL-7R signaling. J Immunol. 2008;181(1):494–502.
Martini S, et al. A critical role for Sox9 in notch-induced astrogliogenesis and stem cell maintenance. Stem Cells. 2013;31(4):741–51.
Miyazaki K, et al. Gene expression profiling of peripheral T-cell lymphoma including gammadelta T-cell lymphoma. Blood. 2009;113(5):1071–4.
Nie L, et al. Notch-induced E2A ubiquitination and degradation are controlled by MAP kinase activities. EMBO J. 2003;22(21):5780–92.
Noguchi A, et al. Zoledronate-activated Vgamma9gammadelta T cell-based immunotherapy is feasible and restores the impairment of gammadelta T cells in patients with solid tumors. Cytotherapy. 2011;13(1):92–7.
Palakodeti A, et al. The molecular basis for modulation of human Vgamma9Vdelta2 T cell responses by CD277/butyrophilin-3 (BTN3A)-specific antibodies. J Biol Chem. 2012;287(39):32780–90.
Pattanapanyasat K. Immune status monitoring of HIV/AIDS patients in resource-limited settings: a review with an emphasis on CD4+ T-lymphocyte determination. Asian Pac J Allergy Immunol. 2012;30(1):11–25.
Plater-Zyberk C, et al. Combined blockade of granulocyte-macrophage colony stimulating factor and interleukin 17 pathways potently suppresses chronic destructive arthritis in a tumour necrosis factor alpha-independent mouse model. Ann Rheum Dis. 2009;68(5):721–8.
Puech PH, et al. Force measurements of TCR/pMHC recognition at T cell surface. PLoS ONE. 2011;6(7):e22344.
Roark CL, et al. Exacerbation of collagen-induced arthritis by oligoclonal, IL-17-producing gamma delta T cells. J Immunol. 2007;179(8):5576–83.

Roux PP, Blenis J. ERK and p38 MAPK-activated protein kinases: a family of protein kinases with diverse biological functions. Microbiol Mol Biol Rev. 2004;68(2):320–44.
Rudolph MG, Stanfield RL, Wilson IA. How TCRs bind MHCs, peptides, and coreceptors. Annu Rev Immunol. 2006;24:419–66.
Spicuglia S, Bonnet M, Ferrier P. Alpha/beta versus gamma/delta T cell development: a choice linked to the transcription factor Sox13. Med Sci (Paris). 2007;23(5):457–8.
Staal FJ, et al. Wnt signaling is required for thymocyte development and activates Tcf-1 mediated transcription. Eur J Immunol. 2001;31(1):285–93.
Sutton C, et al. A crucial role for interleukin (IL)-1 in the induction of IL-17-producing T cells that mediate autoimmune encephalomyelitis. J Exp Med. 2006;203(7):1685–91.
Sutton CE, et al. Interleukin-1 and IL-23 induce innate IL-17 production from gammadelta T cells, amplifying Th17 responses and autoimmunity. Immunity. 2009;31(2):331–41.
Tanigaki K, et al. Regulation of alphabeta/gammadelta T cell lineage commitment and peripheral T cell responses by Notch/RBP-J signaling. Immunity. 2004;20(5):611–22.
Trampont PC, et al. CXCR4 acts as a costimulator during thymic beta-selection. Nat Immunol. 2010;11(2):162–70.
Varnum-Finney B, Brashem-Stein C, Bernstein ID. Combined effects of Notch signaling and cytokines induce a multiple log increase in precursors with lymphoid and myeloid reconstituting ability. Blood. 2003;101(5):1784–9.
Wang S, et al. Anti-interleukin-12/23p40 antibody attenuates chronic rejection of cardiac allografts partly via inhibition gammadeltaT cells. Clin Exp Immunol. 2012;169(3):320–9.
Washburn T, et al. Notch activity influences the alphabeta versus gammadelta T cell lineage decision. Cell. 1997;88(6):833–43.
Welte T, et al. Vgamma4+ T cells regulate host immune response to West Nile virus infection. FEMS Immunol Med Microbiol. 2011;63(2):183–92.
Wrobel P, et al. Lysis of a broad range of epithelial tumour cells by human gamma delta T cells: involvement of NKG2D ligands and T-cell receptor- versus NKG2D-dependent recognition. Scand J Immunol. 2007;66(2–3):320–8.
Zhang ZX, Young K, Zhang L. CD3+CD4-CD8- alphabeta-TCR+ T cell as immune regulatory cell. J Mol Med (Berl). 2001;79(8):419–27.
Zhao N, et al. Vgamma4 gammadelta T cell-derived IL-17A negatively regulates NKT cell function in Con A-induced fulminant hepatitis. J Immunol. 2011;187(10):5007–14.
Zhou J, et al. Anti-gammadelta TCR antibody-expanded gammadelta T cells: a better choice for the adoptive immunotherapy of lymphoid malignancies. Cell Mol Immunol. 2012;9(1):34–44.
Zhu C, et al. Insights from in situ analysis of TCR-pMHC recognition: response of an interaction network. Immunol Rev. 2013;251(1):49–64.
Zocchi MR, et al. High ERp5/ADAM10 expression in lymph node microenvironment and impaired NKG2D ligands recognition in Hodgkin lymphomas. Blood. 2012;119(6):1479–89.

Lingyan Wang graduated from Fudan University in 2009. Currently, she is working as Director of Alive Cell Monitoring Platform at Biomedical Research Center, Zhongshan Hospital, Fudan University. Her main research is on transplant tolerance and immune microenvironment both in acute allograft rejection and tumor development.

Fuping Wang Ms. Wang got her master degree at the Fudan University in 2014, with the finding that the autophagy of peripheral CD4+T cells exists in the progress of chronic HBV infection and can be enhanced by relative risk factors. Currently she is working at the Capital Paediatrics Institute, Peking University. Her main research interest focuses on the immunology of HBV-related liver fibrosis, including the regulation of CD4+T cells on HSCs and other immune cells and their alterations caused by related factors in the liver microenvironment. Ms. Wang has published 4 scientific papers in international well-recognized journals.

Jieyi Shi Dr. Shi got his Ph.D. degree at the Fudan University in 2013, with the findings that margin infiltrating CD20+ B cells display an atypical memory phenotype and correlate with favorable prognosis in hepatocellular carcinoma. Currently he is working at the Liver Cancer Institute, Zhongshan Hospital, Fudan University. His main research interest focuses on tumor immunology, including the regulation of tumor environmental chemotaxis and the functional status of tumor infiltrating lymphocytes. Dr. Shi has published 7 scientific papers in international well-recognized journals.

Xiangdong Wang Dr. Wang works as a distinguished professor of Respiratory Medicine at Fudan University, China, adjunct professor of Molecular Bioscience at North Carolina State University, USA, and adjunct professor of Clinical Bioinformatics at Lund University. He serves as a Director of Biomedical Research Center, Fudan University Zhongshan Hospital. He was also pointed as visiting, guest and honor professors in Harbin Medical University (China), Zhejiang University (China), Kyoto Prefectural University of Medicine (Japan), and National Academy of Medicine (Romania). Dr. Wang acts as an Editor-in-Chief of Clinical & Translational Medicine, Journal of Clinical Bioinformatics and Journal of Epithelial Biology & Pharmacology, and Asian Editor of Journal of Cellular & Molecular Medicine. His main research is focused on the role of epithelial cells in the development of organ dysfunction, disease-specific biomarkers, and potential therapies.

Chapter 8
Prolymphocyte: Lymphocytes and Bioinformatics

Mengjia Qian, Claudio Spada, and Xiangdong Wang

Abstract Prolymphocytes are the immediate precursor of a lymphocyte, derived from a lymphoblast, a developmental form in the lymphocytic series, intermediate between the lymphoblast and lymphocyte.

Pro-B-lymphocytes (pro-B-cells), that develop from hematopoietic progenitor cells, are an early identifiable intermediate cell type in a series of developmental stages leading to the generation of mature B-cells.

Similar to B cell maturation, T cell maturation also involves rearrangements of T cell receptor (TCR) genes and the expression of various membrane markers. Common lymphoid progenitors (CLPs) migrate from the embryonic liver or from adult bone marrow to colonize the thymus, the site of education and maturation of T lymphocyte. Several cytokines, that act together to promote early T-cell development, are also involved in prethymic development, such as the generation of CLPs and the maintenance of hematopoietic stem cells.

Aurora-A kinase is a cell-cycle-regulating kinase responsible for chromosomal segregation. Overexpression of Aurora-A kinase is correlate with tumor proliferation and chromosomal instability, thus Aurora-A kinase overexpression in chronic lymphocytic leukemia (CLL) may be involved in the genesis of chromosomal abnormalities and also is a potential target for therapeutic intervention.

Evidence from recent research indicate the possible relationship between the presence of malignant lymphocytes and their compact nucleoli in patient affected

M. Qian, B.S.
Biomedical Research Center, Zhongshan Hospital, Fudan University, Shanghai, China

Shanghai Clinical Bioinformatics Research Institute, Shanghai, China
e-mail: shilver-s-dream@hotmail.com

C. Spada
Healthcare Accountability Lab, Department of Biomedical Sciences for Health, University of Milan, Milan, Italy
e-mail: cspada@picu.it

X. Wang, Ph.D. (✉)
Zhongshan Hospital, Fudan University, China

Shanghai Institute of Clinical Bioinformatics, China
e-mail: xiangdong.wang@clintransmed.org

X. Wang (ed.), *Single Cell Sequencing and Systems Immunology*,
Translational Bioinformatics 5, DOI 10.1007/978-94-017-9753-5_8

by chronic lymphocytic leukemia. Nucleolar test can be a possible prognostic parameter which can help identify the subset of patients that will run a more progressive course.

B-cell lymphocyte kinase (Blk), B-cell linker protein (BLNK), spleen tyrosine kinase (SYK), zeta-associated protein-70 (ZAP70) and iron-responsive element-binding protein 2 (IREB2) can be identified as important prolymphocyte related proteins that establish systemic interaction.

Cell development is also characterized by specific genes. Thirteen genes included in the Pubmed gene bank are related with prolymphocyte, while others genes play an important role during the maturation of T cells.

CLL is characterized by a level of Prolymphocytes >10 %. Current studies aim to find efficient biomarkers to characterize prognostic and phenotypic features of patients with CLL or prolymphocytic leukemia (PLL).

Although pro-B and pro-T cells have unique characters, both have specific characterization of lymphoid malignancy, and a better understanding of pro-B and pro-T is required. Thus, further studies must be conducted to provide an effective treatment of lymphoid malignancy.

Keywords Prolymphocyte • Proteomics • Genomics

8.1 Introduction

Prolymphocytes are the immediate precursor of a lymphocyte, derived from a lymphoblast, a developmental form in the lymphocytic series, intermediate between the lymphoblast and lymphocyte. The present chapter aims at over-viewing molecular biological and functional characters of these cells, describing cell-specific alterations of cellular functions, phenotypes and mediator productions in pathological conditions, exploring new understanding of cellular mechanisms on basis of findings from genomics, proteomics studies, defining disease-specific biomarkers and targets in these cells and the potential of target-associated therapies as well.

8.2 Cell Biology and Function

The pro-B cells are a population of cells that express $CD34^+$ $CD19^+$ (Nimer 2008). In the development of B cells, successful rearrangement of the immunoglobulin (Ig) heavy (H) chain gene in pro-B cells will result in the surface expression of pre–B cell receptor (pre-BCR), thus resulting in the differentiation to the progenitor B cell stage (Nakayama et al. 2009).

The pro-B cells express Toll-like receptors (TLRs) and respond to TLR ligands (Azulay-Debby et al. 2007) which can stimulate class switch recombination (CSR) in B cells. Developing B cells will undergo CSR in response to mitogenic signals and differentiate to IgG-producing cells, thus indicating that developing B cells have the potential to participate in the immune response (Edry et al. 2008).

The pro-T cells are the earliest stage of the T cell lineage. They are a population of cells that express c-kit^{+}CD44^{+}CD25^{+} while not expressing T cell receptor (TCR), CD3, CD4 or CD8. Recombination activating gene 1 (RAG1) and recombination activating gene 2 (RAG2) are first expressed at this stage.

8.3 Production of Mediators

In the bone marrow, common lymphoid progenitors (CLPs) develop through distinct pro-B cell populations toward mature B cells. Following their activation, B cells secrete various antibodies and cytokines, such as growth factor independent 1 (Gfi1), which is expressed in early B cells, and its expression will decrease when B cells mature (van der Meer et al. 2010).

As similar to B cell maturation, T cell maturation also involves rearrangements of TCR genes and the expression of various membrane markers. CLPs migrate from the embryonic liver or from adult bone marrow to colonize the thymus, the site of education and maturation of T lymphocytes. The commitment to the T cell lineage depends on signals from the thymic microenvironment, where interleukin-7 (IL-7) is indispensable for survival and proliferation of human T cell precursors. Studies also showed that expression of CD127 (the high-affinity chain of the IL-7 receptor, IL-7Ra) in T cell precursors fluctuates prior to T-lineage commitment (Bell and Bhandoola 2008). The proliferation and survival of early T-lineage progenitors also depends on other cytokines such as stem cell factor (SCF) and Flt3 ligand (FL). These data suggest that several cytokines act together to promote early T-cell development. However, these cytokines are also involved in prethymic development, such as the generation of CLPs and the maintenance of hematopoietic stem cells (HSCs) (Thoren et al. 2008).

Aurora-A kinase is a cell-cycle-regulating kinase responsible for chromosomal segregation. Overexpression of this kinase has been shown to correlate with tumor proliferation and chromosomal instability. So Aurora-A kinase overexpression in chronic lymphocytic leukemia (CLL) may be involved in the genesis of chromosomal abnormalities and also is a potential target for therapeutic intervention. Studies showed that the larger prolymphocytes showed stronger Aurora-A kinase expression in CLL (Inamdar et al. 2008).

8.4 Nuclear

The size of prolymphocyte is between 10 and 18 μm. Prolymphocyte has compact nucleolus, prominent nucleoli and condensed nuclear chromatin (Nakashima et al. 2008). There are results that indicate the relationship between the malignant lymphocytes and their compact nucleoli in CLL patients (Klobusicka et al. 2010).

So we think that the nucleolar test can be a possible prognostic parameter which can help identify the subset of patients that will run a more progressive course.

8.5 Proteomics

Proteomics is a biochemical methodology to study proteins expressed in a biological system of interest. Recently, the study of proteomics has proved its importance in investigations on broader aspects (Schultze et al. 2010).

From Pubmed Protein Bank, there are five prolymphocyte related proteins. They are B-cell lymphocyte kinase (Blk), B-cell linker protein (BLNK), spleen tyrosine kinase (SYK), zeta-associated protein-70 (ZAP70) and iron-responsive element-binding protein 2 (IREB2).

SYK plays an important role in the B cell receptor (BCR) signaling, functioning as the major downstream effector protein. Cell signaling initiated by BCR is critical to normal development of B lymphocytes, most notably at the pro-B cell stage. However, inhibition of this signaling pathway with the SYK inhibitor will produce significant efficacy in lymphoid malignancies and autoimmune diseases. Scientists demonstrated that short-term use of SYK inhibitor impaired B lymphocyte development at the pro-B stage without affecting mature B cell populations (Barr et al. 2012).

Signal transduction from the BCR requires recruitment and activation of SYK. Activated SYK phosphorylates several downstream signaling elements, including BLNK (Niiro and Clark 2002). BLNK is a pivotal adapter protein in signal transduction from BCR. The mutation of BLNK may cause a complete block in B cell development at the pro–B cell stage in human (Minegishi et al. 1999).

ZAP70 is a member of the SYK family that is expressed predominantly by T cells and NK cells. It is essential for T cell activation, and B cell development from the pro-B to mature B cells as well (Chen et al. 2002).

Progression through the cell cycle is regulated by sequential events in response to extra and intracellular signals. The cyclin-dependent kinase (CDK) plays an important role in this process. Scientists explored that the expression of cyclin D1 may induce an increase of p27Kip1 (p27) level, which regulates the cell proliferation, in the nucleus and the cytoplasm but not the relocation of the protein in pro-B cells (Bustany et al. 2011).

However, activation of the janus tyrosine kinase 3/signal transducer and activator of transcription 5 (JAK3/STAT5) signaling pathway will lead to the suppression of p27kip1 expression. Activation of the JAK3/STAT5 pathway is partly due to the autocrine production of IL-7 while BLNK inhibits the activation of JAK3 through direct interaction with it and thus suppressing cell-cycle progression and growth (Nakayama et al. 2009).

CCCTC-binding factor (CTCF) is a zinc finger protein that has structural and functional roles in chromatin organization (Phillips and Corces 2009). Scientists have identified that CTCF, most likely in concert with other proteins such as the

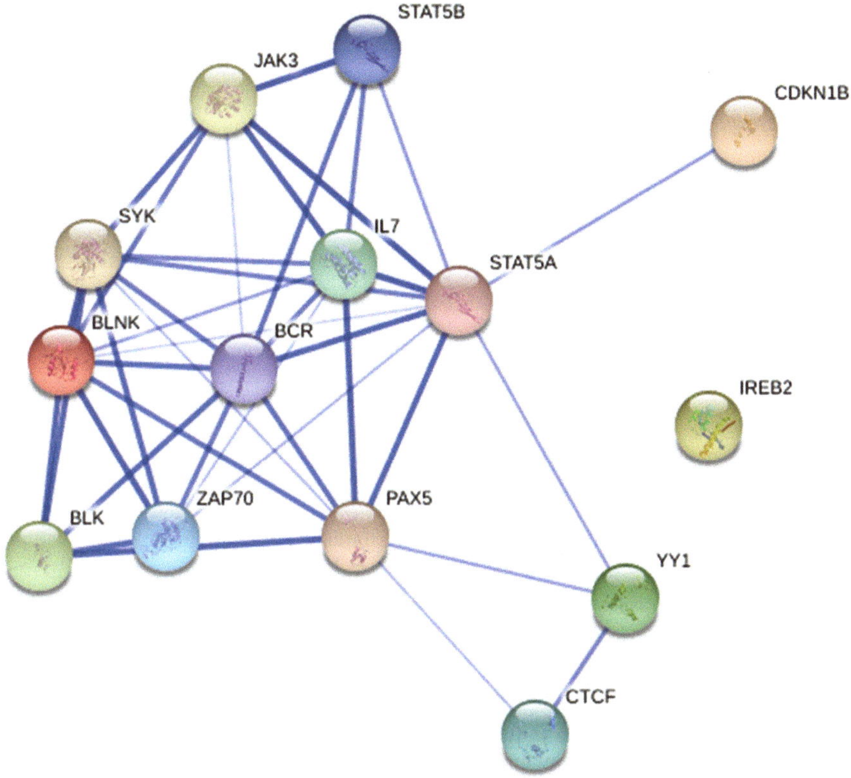

Fig. 8.1 Interaction network on proteomics. Stronger associations are represented by *thicker lines* using the STRING online tool

Yin Yang-1 (YY1) and Paired box 5 (Pax5), is involved in the looping of the Igh locus during the pro–B cell stage, during which the Igh locus undergoes V(D)J rearrangement (Degner et al. 2011).

In summary, we conclude the interaction with these related proteins mentioned above in Fig. 8.1.

8.6 Genomics

From Pubmed Gene bank, there are 13 genes related with prolymphocyte showed in Table 8.1.

Many studies have demonstrated that Notch1 is important in T cell commitment. Notch1 signaling in the thymic cortex instructs the T cell fate and inhibits B cell development. The block to B cell development may be due to the inhibition of EBF function and Pax5 expression by Notch signaling (Smith et al. 2005).

Table 8.1 Genes related with prolymphocyte found in Pubmed

Official symbol	Full name
FLT3	Fms-related tyrosine kinase 3
GATA3	GATA binding protein 3
IL10	Interleukin 10
CXCL12	Chemokine (C-X-C motif) ligand 12
CRP	C-reactive protein, pentraxin-related
NOTCH1	Notch 1
PDCD1	Programmed cell death 1
C3	Complement component 3
IL7	Interleukin 7
PAX5	Paired box 5
Stat5a	Signal transducer and activator of transcription 5A
Myc	Myelocytomatosis oncogene
Il7r	Interleukin 7 receptor

GATA3 likely regulates the differentiation of early T-lineage progenitors, but not the survival or proliferation. Evidence suggested that other transcription factors, such as Ikaros, myeloblastosis oncogene (Myb), and T cell factor-1 (TCF1), also play important roles in the generation and differentiation of the earliest T-lineage progenitors (Yang et al. 2010).

Some other genes not included in the pubmed gene bank also play important roles in the cell development. Homeobox B3 (HOXB3) expresses in low immature progenitors and probably interferes with the choice of sub-lineage in pro-T cells. Homeobox C4 (HOXC4) is expressed at all stages of T cell development (Taghon et al. 2003).

As a thymus-specific noncoding RNA, Thy-ncR1 gene specifically expresses during the maturation of T cells. Thy-ncR1 is required to explore the nuclear functions of this transcript, so it is a functional cytoplasmic “riboregulator” at the maturation of T cells (Aoki et al. 2010).

The signaling pathway through which Flt3 promotes the generation of B lineage progenitors is unknown, but in vitro data suggest that activation of this receptor promotes expression of the IL-7 receptor (IL-7R). The development of pro-B cells in the bone marrow requires signaling through IL-7R. Furthermore, in a culture system, IL-7R signaling is sufficient to induce the differentiation of CLPs into pro-B cells. Importantly, two studies have shown that combined loss of Flt3 and IL-7R signaling results in a complete failure to develop B lineage cells during both fetal and adult hematopoiesis (Laslo et al. 2008).

IL-7R signaling may directly regulate immunoglobulin gene rearrangement via STAT5 at the pro-B cell stage. Stimulation of IL-7 is important in maintaining EBF expression, a transcription factor indispensable for B cell development, after B cell lineage commitment in pro-B cells. It is critical for maintaining B cell differentiation potential and progression of pro-B cells (Kondo 2010). Another study showed other two genes, E47 and Pax5, related with the transcription of EBF (Northrup and Allman 2008).

Stromal-derived factor 1 (SDF1) plays an important role in B lymphopoiesis. Scientists found that mice lack SDF1 or its receptor, chemokine CXC motif receptor 4 (CXCR4), have severely reduced the number of B cells (Tokoyoda et al. 2004). Theses results suggest that the SDF1/CXCR4 axis regulates the localization of pro-B cells into the specific bone marrow microenvironment. Increase of tumor necrosis factor alpha (TNF-α) will inhibit the production of SDF1 from bone marrow stromal cells, suggesting that TNF-α might change the bone marrow microenvironment and may directly or indirectly suppress the expansion of developing B cells (Ueda et al. 2004).

Nfil3 (nuclear factor IL-3), also known as E4bp4 (E4-binding protein 4), is a basic leucine zipper (bZIP) transcriptional activator and was initially described for its ability to limit viral transcription and for promoting IL-3-mediated survival of pro-B cells (Klein et al. 2010).

The phosphatase and tensin homolog gene (PTEN) on chromosome 10 controls the first step in the phosphatidylinositol 3 kinase (PI3K)-Akt-mammalian target of rapamycin (mTOR) pathway. Scientists have identified that intracellular Pten level in pro-B cells leads to the low AKT activity and is responsible to the cell death (Cheng et al. 2009).

In summary, we conclude the interaction with these related genes mentioned above in Fig. 8.2.

8.7 Disease-Specific Biomarkers

In Chronic lymphocytic leukaemia, >10 % prolymphocytes is characterized to this disease. The tumor suppressor protein p53 is an important transcription factor in the cellular response to DNA damage and may activate apoptotic pathways (Lu et al. 2010). Scientists have verified the important role of the p53 gene in prolymphocytes in CLL and its possible involvement in the pathogenesis of a progressive form of CLL (Lens et al. 1997).

Furthermore, scientists found that two genes, B-cell lymphoma 11A (BCL11A) and IgH may related with the t(2;14) in CLL, which may represent a distinct clinic pathologic entity characterized by atypical morphologic features (Yin et al. 2009).

Current studies aim to find more efficient biomarkers to characterize prognostic and phenotypic features of patients with CLL or prolymphocytic leukemia (PLL).

Myelocytomatosis oncogene (MYC) is the prototype for oncogene activation by chromosomal translocation (Boxer and Dang 2001). MYC rearrangement results in overexpression of MYC protein and induces cell proliferation. MYC rearrangement occurs rarely in CLL prolymphocytes, but is associated with increased prolymphocyte numbers, complex cytogenetic abnormalities, and a poor prognosis (Huh et al. 2008).

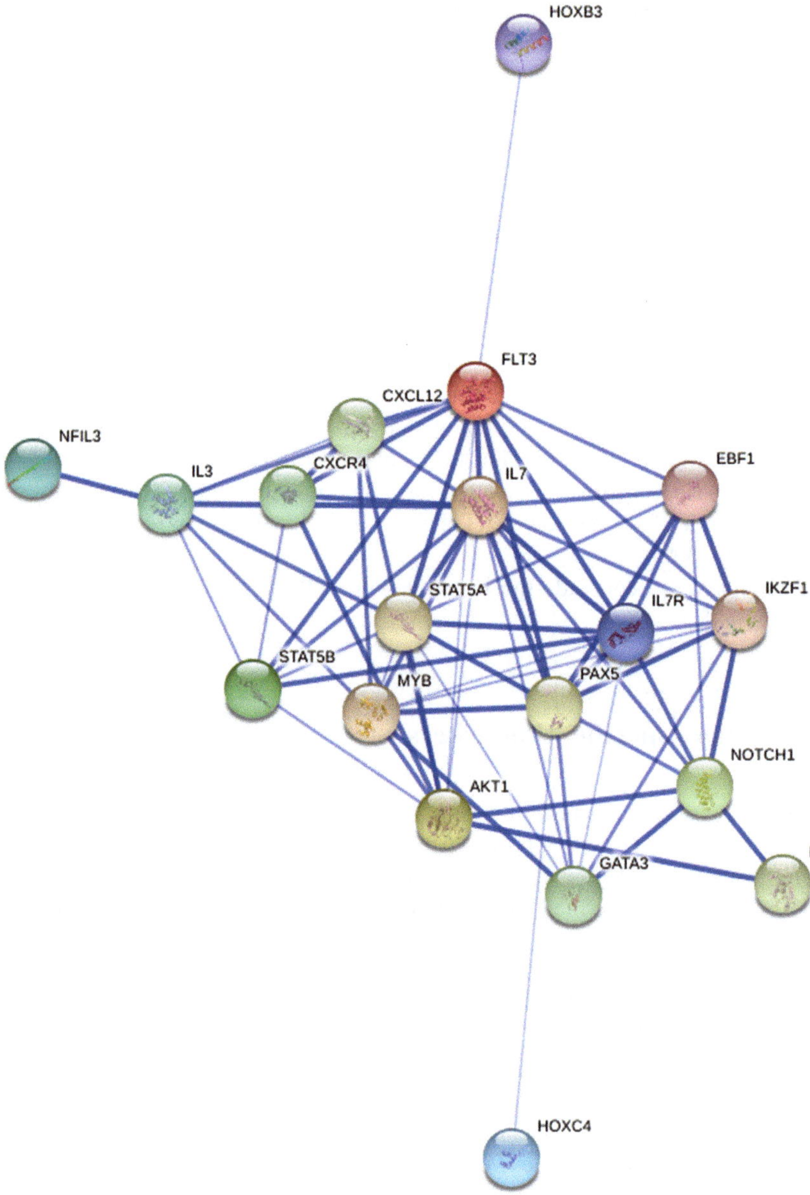

Fig. 8.2 Interaction network on genomics. Stronger associations are represented by *thicker lines* using the STRING online tool

Since MYC translocations seem to be associated with poor survival in CLL and PLL, the translocation of MYC may play an important role in bringing the pathogenesis and progression of CLL or PLL into a more aggressive diseases (Put et al. 2012).

8.8 Roles in the Diseases

In Chronic lymphocytic leukaemia, >10 % prolymphocytes is characterized to this disease. The observation that the leukemic cells in most children with acute lymphoblastic leukemia were classified as pro-B cells, suggesting the importance of this developmental stage (Martensson et al. 2010).

As to old people, a study described patients aged 62–79 with splenomegaly and bone marrow involvement by splenic B-cell lymphoma who developed more than 55 % prolymphocytes in blood. The diagnosis of B cell prolymphocytic leukemia was considered clinically based on a markedly elevated prolymphocyte count (Hoehn et al. 2012). Also the unusual increase of pro-T cells can cause T-cell prolymphocytic leukemia (Ozaki et al. 1992).

8.9 Future/Perspective

In this chapter, we have described about the function and the biology of prolymphocytes, the mediators generated by them, the proteomics, genomics of them and their important role played in diseases as well. Though pro-B and pro-T cells have unique characters, they are all specific characterization of lymphoid malignancy.

We still have a long way to further understand these cells, until we can well understand the lymphoid malignancy and other related diseases, thus finding effective ways to treat them.

References

Aoki K, et al. A thymus-specific noncoding RNA, Thy-ncR1, is a cytoplasmic riboregulator of MFAP4 mRNA in immature T-cell lines. BMC Mol Biol. 2010;11:99.

Azulay-Debby H, Edry E, Melamed D. CpG DNA stimulates autoreactive immature B cells in the bone marrow. Eur J Immunol. 2007;37(6):1463–75.

Barr PM, et al. Syk inhibition with fostamatinib leads to transitional B lymphocyte depletion. Clin Immunol. 2012;142(3):237–42.

Bell JJ, Bhandoola A. The earliest thymic progenitors for T cells possess myeloid lineage potential. Nature. 2008;452(7188):764–7.

Boxer LM, Dang CV. Translocations involving c-myc and c-myc function. Oncogene. 2001;20(40):5595–610.

Bustany S, Tchakarska G, Sola B. Cyclin D1 regulates p27(Kip1) stability in B cells. Cell Signal. 2011;23(1):171–9.
Chen L, et al. Expression of ZAP-70 is associated with increased B-cell receptor signaling in chronic lymphocytic leukemia. Blood. 2002;100(13):4609–14.
Cheng S, et al. BCR-mediated apoptosis associated with negative selection of immature B cells is selectively dependent on Pten. Cell Res. 2009;19(2):196–207.
Degner SC, et al. CCCTC-binding factor (CTCF) and cohesin influence the genomic architecture of the Igh locus and antisense transcription in pro-B cells. Proc Natl Acad Sci U S A. 2011;108(23):9566–71.
Edry E, Azulay-Debby H, Melamed D. TOLL-like receptor ligands stimulate aberrant class switch recombination in early B cell precursors. Int Immunol. 2008;20(12):1575–85.
Hoehn D, et al. Splenic B-cell lymphomas with more than 55% prolymphocytes in blood: evidence for prolymphocytoid transformation. Hum Pathol. 2012;43(11):1828–38.
Huh YO, et al. MYC translocation in chronic lymphocytic leukaemia is associated with increased prolymphocytes and a poor prognosis. Br J Haematol. 2008;142(1):36–44.
Inamdar KV, et al. Aurora-A kinase nuclear expression in chronic lymphocytic leukemia. Mod Pathol. 2008;21(12):1428–35.
Klein WR, et al. The intrathymic crossroads of T and NK cell differentiation. Immunol Rev. 2010;238(1):126–37.
Klobusicka M, et al. Possible prognostic value of nucleolar morphology in pathologic cells of B-chronic lymphocytic leukemia. Neoplasma. 2010;57(5):429–37.
Kondo M. Lymphoid and myeloid lineage commitment in multipotent hematopoietic progenitors. Immunol Rev. 2010;238(1):37–46.
Laslo P, et al. Gene regulatory networks directing myeloid and lymphoid cell fates within the immune system. Semin Immunol. 2008;20(4):228–35.
Lens D, et al. p53 abnormalities in CLL are associated with excess of prolymphocytes and poor prognosis. Br J Haematol. 1997;99(4):848–57.
Lu TP, et al. Distinct signaling pathways after higher or lower doses of radiation in three closely related human lymphoblast cell lines. Int J Radiat Oncol Biol Phys. 2010;76(1):212–19.
Martensson IL, et al. The pre-B cell receptor checkpoint. FEBS Lett. 2010;584(12):2572–9.
Minegishi Y, et al. An essential role for BLNK in human B cell development. Science. 1999;286(5446):1954–7.
Nakashima H, et al. Splenic irradiation as a successful treatment for an elderly patient with B-cell prolymphocytic leukemia. Rinsho Ketsueki. 2008;49(12):1619–22.
Nakayama J, et al. BLNK suppresses pre-B-cell leukemogenesis through inhibition of JAK3. Blood. 2009;113(7):1483–92.
Niiro H, Clark EA. Regulation of B-cell fate by antigen-receptor signals. Nat Rev Immunol. 2002;2(12):945–56.
Nimer SD. MDS: a stem cell disorder – but what exactly is wrong with the primitive hematopoietic cells in this disease? Hematol Am Soc Hematol Educ Program. 2008;2008:43–51.
Northrup DL, Allman D. Transcriptional regulation of early B cell development. Immunol Res. 2008;42(1–3):106–17.
Ozaki S, et al. A case of T-cell prolymphocytic leukemia. Rinsho Ketsueki. 1992;33(6):817–22.
Phillips JE, Corces VG. CTCF: master weaver of the genome. Cell. 2009;137(7):1194–211.
Put N, et al. Chronic lymphocytic leukemia and prolymphocytic leukemia with MYC translocations: a subgroup with an aggressive disease course. Ann Hematol. 2012;91(6):863–73.
Schultze FC, et al. Differential proteome and phosphoproteome signatures in human T-lymphoblast cells induced by sirolimus. Cell Prolif. 2010;43(4):396–404.
Smith EM, et al. Inhibition of EBF function by active Notch signaling reveals a novel regulatory pathway in early B-cell development. Blood. 2005;106(6):1995–2001.
Taghon T, et al. Homeobox gene expression profile in human hematopoietic multipotent stem cells and T-cell progenitors: implications for human T-cell development. Leukemia. 2003;17(6):1157–63.

Thoren LA, et al. Kit regulates maintenance of quiescent hematopoietic stem cells. J Immunol. 2008;180(4):2045–53.
Tokoyoda K, et al. Cellular niches controlling B lymphocyte behavior within bone marrow during development. Immunity. 2004;20(6):707–18.
Ueda Y, et al. Inflammation controls B lymphopoiesis by regulating chemokine CXCL12 expression. J Exp Med. 2004;199(1):47–58.
van der Meer LT, Jansen JH, van der Reijden BA. Gfi1 and Gfi1b: key regulators of hematopoiesis. Leukemia. 2010;24(11):1834–43.
Yang Q, Jeremiah BJ, Bhandoola A. T-cell lineage determination. Immunol Rev. 2010;238(1): 12–22.
Yin CC, et al. Chronic lymphocytic leukemia With t(2;14)(p16;q32) involves the BCL11A and IgH genes and is associated with atypical morphologic features and unmutated IgVH genes. Am J Clin Pathol. 2009;131(5):663–70.

Mengjia Qian is working as a research scientist in biomedical research center, Zhongshan Hospital, Fudan University. She is currently working on the next generation sequencing field. Her research is mainly focused on immunology and has published 7 scientific papers. Email: shilver-s-dream@hotmail.com

Claudio Spada serves as President of the Scientific Society Healthcare Safety Accountability and is Associated for Research Programs to the Healthcare Accountability Lab, Department of Biomedical Sciences for Health, University of Milan, Italy and to The First Hospital of Wenzhou Medical University, Wenzhou, Zhenjiang, China. He has long been engaged in research work on biomedicine and bioinformatics. Email: cspada@picu.it

Xiangdong Wang Dr. Wang works as a distinguished professor of Respiratory Medicine at Fudan University, China, adjunct professor of Molecular Bioscience at North Carolina State University, USA, and adjunct professor of Clinical Bioinformatics at Lund University. He serves as a Director of Biomedical Research Center, Fudan University Zhongshan Hospital. He was also pointed as visiting, guest and honor professors in Harbin Medical University (China), Zhejiang University (China), Kyoto Prefectural University of Medicine (Japan), and National Academy of Medicine (Romania). Dr. Wang acts as an Editor-in-Chief of Clinical & Translational Medicine, Journal of Clinical Bioinformatics and Journal of Epithelial Biology & Pharmacology, and Asian Editor of Journal of Cellular & Molecular Medicine. His main research is focused on the role of epithelial cells in the development of organ dysfunction, disease-specific biomarkers, and potential therapies.

Chapter 9
Lymphocytes in Liver Cancer

Xiaojing Xu, William C.S. Cho, and Jinglin Xia

Abstract Tumor infiltrating lymphocytes (TIL) are an important part of tumor surveillance systems and a representative component of antitumor immune responses. The liver, as an immunological organ, is specially equipped with liver-associated lymphocytes. Liver cancer can be regarded as an unresolved lesion of chronic inflammation and its microenvironment can control tumor progression via TIL activation and proliferation, predominantly CD8(+) and CD4(+) T lymphocytes and natural killer cells. However, increased CD4(+) CD25(+) Foxp3(+) T-regulatory lymphocytes (Tregs) may impair the effector function of CD8(+) T cells, thus promote liver cancer progression and invasion. Moreover, the intratumoral balance of regulatory and cytotoxic T cells is a promising independent predictor for recurrence and survival, which indicates combination of the depletion of Tregs and the concomitant stimulation of effector T cells may be an effective immunotherapy strategy for liver cancer. As an emerging discipline, translational bioinformatics specially focus on genomics, proteomics, metabolomics, and bioinformatics. It can increase our understanding in the molecular mechanisms of systems immunology. This chapter aims to overview the biological and functional characters of different lymphocytes in liver cancer, describe lymphocyte-related cytokines, chemokines, mediators, lymphocyte-related bioinformatics and network in order to define liver cancer specific biomarkers as well as therapeutic targets.

Keywords Lymphocyte • Liver cancer • T-regulatory lymphocytes (Tregs) • Tumor-infiltrating lymphocyte (TIL) • Immunotherapy

X. Xu, M.D. • J. Xia, M.D., Ph.D. (✉)
Department of Medical Oncology, Fudan University, Zhongshan Hospital, Shanghai, China
e-mail: xiajinglin@fudan.edu.cn

W.C.S. Cho
Department of Clinical Oncology, Queen Elizabeth Hospital, Kowloon, Hong Kong

X. Wang (ed.), *Single Cell Sequencing and Systems Immunology*,
Translational Bioinformatics 5, DOI 10.1007/978-94-017-9753-5_9

9.1 Introduction

Liver cancer is the fifth most common cancer worldwide and the second most frequent cause of cancer-related death, especially in East and South-East Asia as well as in Middle and Western Africa (Jemal et al. 2011). Among primary liver cancers, hepatocellular carcinoma (HCC) represents the major histological subtype that accounting for 70–85 % (Perz et al. 2006), its incidence has significantly increased in the last two decades in close relation with hepatitis B virus (HBV) and hepatitis C virus (HCV), aflatoxin and alcohol abuse (Forner et al. 2012; Patel et al. 2012; Revill and Yuan 2013; Maini and Schurich 2010; Lan et al. 2013). The only effective treatment approaches for patients with liver cancer are resection or liver transplantation. Following transplantation, there is an 85 % 5-year survival and less than 10 % recurrence rate in highly selected patients (single tumor <5 cm in diameter or fewer than three tumors <3 cm each in diameter) (Silva and Sherman 2011). However, the majorities of patients can not meet such strict criteria or have other contraindications, who can accept local therapies such as percutaneous ethanol injection (PEI), thermal ablation or transarterial chemoembolization (TACE) (Patel et al. 2012).

Although survival of patients with liver cancer has improved after decades of research, it is still devastating due to the high rate of recurrence and metastasis. Hepatocarcinogenesis and metastasis are haphazard, stepwise processes involving multiple regulatory network pathways, adaptation, avoidance of immune surveillance, and finally emergence of properties that allow disseminated growth, metastasis, and typically the ultimate death (Forner et al. 2012). Among them, immunological mechanisms are important in the surveillance of malignancy and control of tumor progression. The liver, as an immuneprivileged organ, especially harbor with liver-associated lymphocytes, mainly T lymphocytes and natural killer (NK) cells, has a pivotal role in immune regulation locally and systemically (Starzl 2001). The mechanisms for HBV persistence and the pathogenesis of chronic HBV have been shown to be mainly due to defects in host immune responses, including defects in T cytolytic cells, T helper cells, regulatory T cells, NK cells, NKT cells, dendritic cells, imbalance in cytokine responses, disturbance in apoptosis and PD-1 responses (Revill and Yuan 2013; Maini and Schurich 2010; Lan et al. 2013; Buchmann et al. 2013; Ren et al. 2013; Huang et al. 2012).

In the liver cancer microenvironment, tumor infiltrating lymphocytes (TIL) represent the host immune response, which is thought to be a result of the changes of surface components of tumor cells. The innate as well as the adaptive immune system is involved in tumor destruction with cell-mediated mechanisms. Besides, liver cancer can be regarded as an unresolved lesion of chronic inflammation and its microenvironment can control tumor progression via TIL activation and proliferation, predominantly CD8(+) and CD4(+) T lymphocytes and NK cells. However, increased CD4(+) CD25(+) Foxp3(+) T-regulatory lymphocytes (Tregs)

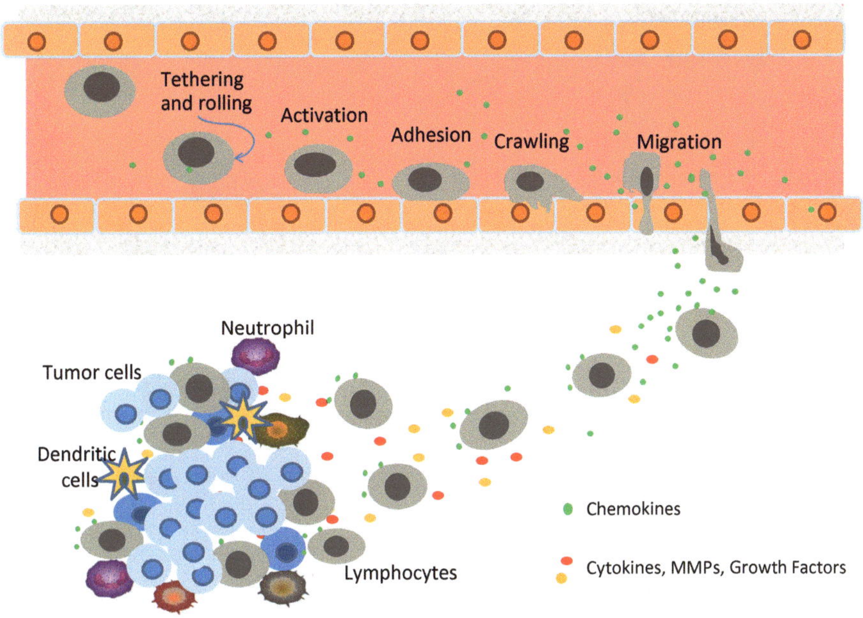

Fig. 9.1 Tumor infiltrating lymphocytes can be recruited to the tumor place by several steps, including tethering and rolling, activation, firm adhesion, crawling, and migration. The tumor microenvironment is constituted of tumor cells, immune cells, cytokines, chemokines, and so on

may impair the effector function of CD8(+) T cells, thus promote liver cancer progression and invasion (Fig. 9.1). Moreover, the intratumoral balance of regulatory and cytotoxic T cells is a promising independent predictor for recurrence and survival, which indicates combination of the depletion of Tregs and the concomitant stimulation of effector T cells may be an effective immunotherapy strategy for liver cancer. The present chapter aims to overview the biological and functional characters of different lymphocytes in liver cancer, describe lymphocyte-related cytokines, chemokines and other mediators, in order to define liver cancer specific biomarkers and therapeutic targets.

9.2 TIL Subsets in Liver Cancer

TILs are composed of different lymphocyte subsets. The pro-tumor or anti-tumor characteristics of TIL are dependent of their component lymphocyte subsets, which are determined by the tumor microenvironment. The following subsets of lymphocytes are found as TILs in liver cancer (Fig. 9.2).

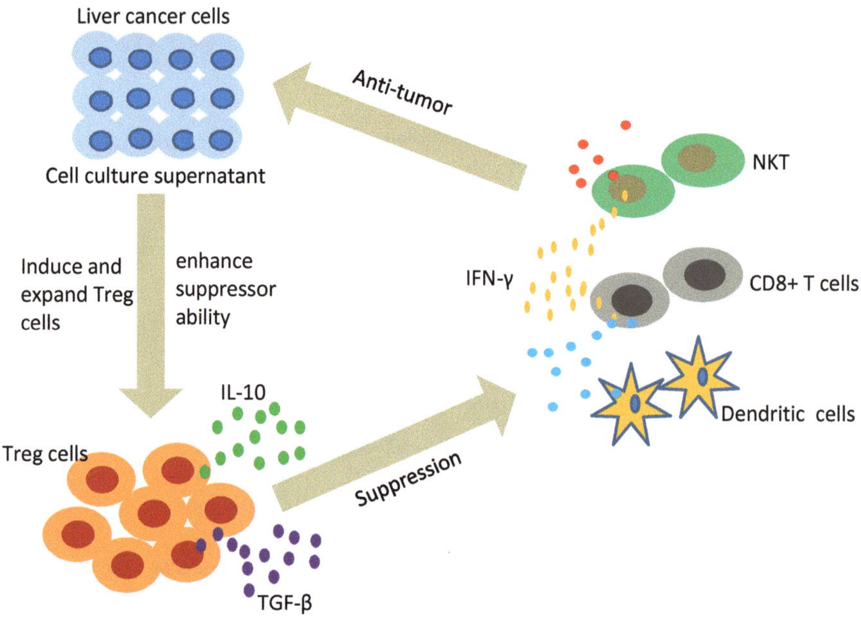

Fig. 9.2 Functions of tumor infiltrating lymphocytes in liver cancer

9.2.1 CD8(+) T Lymphocytes

CD8(+) T cells are a crucial component of antitumor immunity. As tumor antigen-specific cytotoxic T lymphocytes that can directly kill target cells by releasing granules including membrane-lytic materials (such as perforin and granzymes in acquired immune responses), they thereby play a central role in antitumor immunity. In liver cancer, extremely marked infiltration of T cells including predominant CD8(+) T cells have been shown to be closely associated with low recurrence and good prognosis (Wada et al. 1998). Furthermore, CD8(+) T cells specific for the self-antigen AFP are present in the normal T cell repertoire and are not centrally or peripherally deleted, supporting the strategies to boost AFP-specific CD8(+) T cell responses in patients with HCC but also demonstrating a diversity of immune responses that may be needed for protection (Thimme et al. 2008). IFN-gamma produced by interleukin (IL)-12-activated tumor-infiltrating CD8(+) T cells directly induce apoptosis of HCC cells (Komita et al. 2006).

However, IL-17- producing CD8(+) T cells are accumulated in human HCC, where they promote disease progression by fostering angiogenesis. Tumor-activated monocytes secreted a set of key cytokines (IL-1 beta, IL-6, and IL-23) to trigger the proliferation of proinflammatory IL-17-producing CD8(+) T cells (Tc17 cells), revealing an intriguing mechanism in which human Tc17 cells are generated by a fine-tuned collaboration among different types of immune cells in distinct tumor microenvironments (Kuang et al. 2010).

9.2.2 Natural Killer T Lymphocytes (NKT)

NKT cells are T cells that serve as a bridge between the innate and adaptive immune system, acting as first responders (Godfrey et al. 2012). In cancer, NKT cells contribute to anti-tumor immunity or suppressing it (Smyth and Godfrey 2000). The protective NKT cells were found to be primarily type I NKT cells defined by the use of a semiinvariant T cell receptor involving Vα14Jα18 in mice and Vα24Jα18 in humans and responding to α-galactosylceramide, and the most protective were among the minority that are CD4(−) (Crowe et al. 2005). The protective activity was dependent on the production of IFN-γ and on the activation of NK cells, but not on perforin in the NKT cells themselves, suggesting that the protective mechanism was indirect, involving IFN-γ-mediated activation of NK cells and possibly CD8(+) T cells that lyse the tumor cells, rather than direct lysis by NKT cells (Smyth et al. 2002). The protective activity was especially mediated by a subset of type I NKT cells found in the liver, rather than the spleen or thymus, and that were negative for CD4 and CD8 (Crowe et al. 2005). The suppressive NKT cells were found to be CD4(+) and to be primarily type II NKT cells that have diverse T cell receptors and respond to other lipids. Furthermore, the type I and type II NKT cells were found to counter-regulate each other, forming a new immunoregulatory axis. The balance along this axis can affect immunity to tumors and responses to vaccines (Berzofsky and Terabe 2009).

Liver-associated NKTs are a unique subset of lymphocytes that play a role in anti-tumor immunity and the mediate suppression of HCC growth via tumor antigen-pulsed dendritic cells (DCs) (Shibolet et al. 2003). The decrease of NKT cells in cirrhotic livers with HCV may involve in their susceptibility to HCC (Kawarabayashi et al. 2000). The viability and function of NKT cells are impaired by CCL20-producing TAMs serving as a hypoxic trap, which can be reversed by adoptive immunotherapy with IL-15-transduced NKT cells (Liu et al. 2012), because IL-15 protects NKT cells from inhibition by tumor-associated macrophages and enhances anti-metastatic activity.

A tumor cell vaccine incorporating alpha-galactosylceramide that targets the immune adjuvant properties of NKT cells is sufficient to significantly inhibit the growth of established tumors and prolong survival in several hematopoietic tumor models. It is also therapeutically effective against acute myeloid leukemia, demonstrating a potent immune adjuvant effect of NKT cell ligands in therapeutic anticancer vaccination and supporting clinical investigation of NKT cell-based immunotherapy in patients with cancers (Mattarollo et al. 2012).

9.2.3 Lymphoid DCs

DCs are the most potent professional antigen-presenting cells, widely distributed in peripheral lymphatic tissues. DCs determine whether an immune response finally activates immune system or leads to immunological tolerance (Kato et al. 2011; Hirohashi and Sato 2011). Although the significance of DCs on the immune regulation of liver is well established, DCs' role in the occurrence and progression of HCC is still uncertain. Marked infiltration of DCs in HCC nodules is closely related to the prognosis of HCC after surgical resection and can be served as a predictive index for recurrence and metastasis of HCC (Cai et al. 2006).

9.2.4 T-Regulatory Lymphocytes (Tregs)

Tregs are a subgroup of CD4(+) T cells with suppressor function characterized by expression of the interleukin 2 receptor α chain (CD25) and Foxp3 (Tanchot et al. 2013). Tregs secrete transforming growth factor (TGF)-β1 and IL-10 (Jonuleit et al. 2001) and may express membrane-bound TGF-β1, the levels of which are increased upon stimulation *in vitro* (Green et al. 2003). In addition, Tregs confer protection against autoimmunity and mediate transplant tolerance (Shalev et al. 2012) and they may play a role in the regulation of tumor immunity (Pedroza-Gonzalez et al. 2013). Increased number of Tregs is present within TILs of non–small cell lung, ovarian, breast, and pancreatic cancers (Liyanage et al. 2002). Local immune responses in patients with HCC are also held in check by tumor-infiltrating Tregs (Wang et al. 2012), which inhibit the activation, proliferation, degranulation and production of granzyme A, granzyme B and perforin of effector CD4(+) and CD8(+) T cells (Fu et al. 2007), partly through inhibition of IL-2 production (Gasteiger and Kastenmuller 2012).

The prevalence of Tregs significantly increases in HCC than in the nontumorous liver in a stepwise manner and that of CD8(+) T cells decreases during the progression of hepatocarcinogenesis (Kobayashi et al. 2007). Besides, the intratumoral balance of regulatory and cytotoxic T cells is associated with prognosis of HCC after resection (Gao et al. 2007). The presence of low intratumoral Tregs in combination with high intratumoral activated CD8+ cytotoxic cells (CTLs) is an independent prognostic factor for both improved disease-free survival (DFS) and overall survival (OS) (Gao et al. 2007). Five-year OS and DFS rates are only 24.1 % and 19.8 % for the group with intratumoral high Tregs and low activated CTLs respectively, comparing with 64.0 % and 59.4 % for the group with intratumoral low Tregs and high activated CTLs, respectively (Gao et al. 2007). In addition, high Tregs density is associated with the absence of tumor encapsulation and the presence of tumor vascular invasion (Gao et al. 2007).

Tregs are also increased in the peripheral blood mononuclear cells (PBMCs) from HCC patients and positively correlate with tumor burden (Cao et al. 2007a). When PBMCs are co-cultured with liver cancer cell lines, Tregs increase in frequency and undergo phenotypic and functional changes (Cao et al. 2007a). CD45RA, CD45RO, CD69, CD62L, GITR, CTLA-4, Ki67, granzyme A, granzyme B, and FOXP3 expressions are upregulated in Tregs after *in vitro* exposure to HCC cell lines resulting in higher suppressor ability of Tregs, indicating that tumor-related factors not only induce and expand Tregs but also enhance their suppressor ability (Cao et al. 2007a).

The intratumoral prevalence of Foxp3(+) Tregs is also associated with a high density of macrophages. The depletion of tissue macrophages attenuates the increase of liver Foxp3(+) Tregs frequency attributed to *in vivo* inoculation with hepatoma, whereas macrophages exposed to tumor culture supernatants from hepatoma-derived cell lines increase Foxp3(+) Tregs frequency *in vitro*. This increase is partially blocked by anti-IL-10 antibody. Thus tumor-associated macrophages may trigger a rise of the intratumoral Foxp3(+) Tregs population, which in turn may promote HCC progression (Zhou et al. 2009). Moreover, CD4(+) CD25(+) Foxp3(+) Tregs can be induced by CD14(+)HLA-DR-/low cells, a new population of myeloid-derived suppressor cells (MDSC) increase in the blood and tumor of HCC patients and exert immunosuppressive function (Hoechst et al. 2008). When Tregs are added to the mixed lymphocyte reaction activated by autologous tumor lysate-pulsed DCs, the proliferation of lymphocytes was inhibited, indicating that these Tregs appear to suppress the immune response activated by DCs (Lee et al. 2012).

9.3 Therapy Based on TIL

In 1986, Rosenberg et al. (Rosenberg et al. 1986) claimed that the combination of TIL and cyclophosphamide mediated the elimination of large metastatic cancer deposits in the liver and lung, which was further potentiated by the simultaneous administration of IL-2. With the combination of cyclophosphamide, TIL and IL-2, 100 % of mice bearing the MC-38 colon adenocarcinoma were cured of advanced hepatic metastases, and up to 50 % of mice were cured of advanced pulmonary metastases. Since then, immunotherapy based on TIL developed rapidly.

9.3.1 Autologous TIL

Adoptive cell therapy (ACT) using autologous TIL is a promising treatment for cancer unresponsive to conventional therapies (Fig. 9.3). An ongoing Phase II clinical trial testing the efficacy of ACT using TIL in metastatic melanoma patients showed that 15/31 (48.4 %) patients had an objective clinical response using immune-related response criteria, with two patients (6.5 %) having a complete response. Progression-free survival of >12 months was observed for 9/15 (60 %) of the responding patients. Factors significantly associated with objective tumor regression included a higher number of TIL infused, a higher proportion of CD8(+) T cells in the infusion product, a more differentiated effector phenotype of the CD8(+)

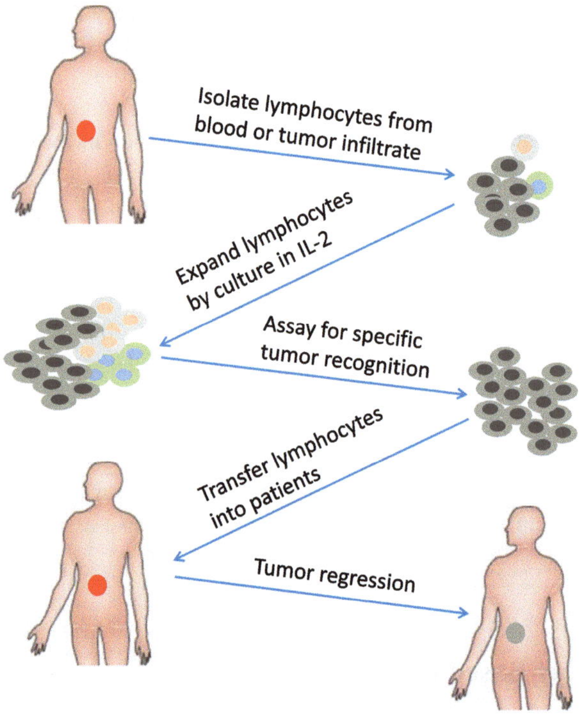

Fig. 9.3 Adoptive cell therapy using autologous tumor infiltrating lymphocytes

population and a higher frequency of CD8(+) T cells co-expressing the negative costimulation molecule "B and T lymphocyte attenuator" (BTLA), indicating that immunotherapy with expanded autologous TIL is capable of achieving durable clinical responses in metastatic melanoma patients and that CD8(+) T cells in the infused TIL, particularly differentiated effectors cells and cells expressing BTLA, are associated with tumor regression (Radvanyi et al. 2012). Clinical studies also indicated that autologous cytokine-induced killer (CIK) cell therapy could be used as an efficient adjuvant anticancer immunotherapy to eradicate residual cancer cells, prevent recurrence, improve progression-free survival (PFS) rates, and promote the quality of life for HCC patients. Comparing T-lymphocyte subsets in peripheral blood, Yue et al. showed the percentage of CD3(+), CD4(+), CD4(+) CD8(+) and CD3(+) CD4(+) T cells significantly increased in the peripheral blood of the CIK group, compared with the non-CIK group (Ma et al. 2012).

Clinical trials on ACT using expanded TIL have shown response in liver cancer, while the successful application of this therapy requires the selection of unique tumor-reactive lymphocyte cultures for each patient. This is a technically and logistically difficult undertaking, and patients who do not have tumor-reactive TIL are not considered eligible for treatment (Prieto et al. 2010). Parameters such as age, sex, type and timing of prior systemic therapy significantly affect the success rate of the initial TIL outgrowth from tumor fragments for ACT; these parameters may be helpful in selecting patients for ACT (Joseph et al. 2011).

9.3.2 *Vaccine*

Tumor vaccine immunotherapy is a promising approach for cancer treatment that is based on stimulation of patients' immune response, and using peptides as therapeutic agents is useful in both prevention of cancer and treatment of established tumors (Marchesi 2013).

The T helper 1 (Th1) immune reaction is very important in DC-based immunotherapy. IL-18, a Th1-biasing cytokine, plays a pivotal role in inducing cytotoxic T lymphocyte responses. A vaccination therapy using DC co-transduced with the tumour-associated antigen (TAA) and IL-18 genes is an effective strategy for immunotherapy in terms of the activation of DCs, CD4(+) T cells and CD8(+) T cells, and may be useful in the clinical application for liver cancer vaccine (Cao et al. 2007b).

9.3.3 *Virus-Specific T Cells*

T cell receptor re-directed HBV-specific T cells generated from PBMC of chronic HBV and HBV-related HCC patients are multifunctional and capable of recognizing HBV-infected cells and HCC tumor cells expressing viral antigens from naturally integrated HBV DNA (Gehring et al. 2011). These genetically modified T cells can be used to reconstitute virus-specific T cell immunity in chronic HBV patients and target tumors in HBV-related HCC.

9.4 Researches Using High-Throughput Technologies on the Lymphocytes in Liver Cancer

9.4.1 *Genomics*

In genomic era, immunologists are ready for monitoring many different gene products simultaneously. It is a common challenge to remember what CD-one-hundred-and-something is, and an ever-increasing number of colours are required for identification on the flow cytometer. Gene expression arrays now offer the possibility of extending this approach beyond the cell surface and expanding it dramatically to survey the entire catalogue of gene transcripts in a lymphoid cell

(Glynne et al. 2000). Lymphocyte-specific genomic instability is a force that links evolution, development and cancer. Kirsch et al. have developed a quantifiable assay for a particular kind of lymphocyte-specific genetic instability (antigen receptor trans-rearrangements) mediated by the V(D)J recombinase complex and find that the level of this type of instability correlates (at the population level) with risk for lymphoid malignancy (Kirsch and Lista 1997). High-resolution genomic profiling of cancers reveals novel recurrent genetic lesions affecting pathways involved in lymphocyte differentiation and cell cycle progression, such as CDKN2A, CDKN1B and RB1 controlling G1/S cell cycle progression, followed by genes associated with B-cell development (Kuiper et al. 2007).

In 2000, Li et al. presented a rapid and high throughput screening method for analyzing the pattern of gene expression of tumor-infiltrating lymphocyte (TIL), which can minimize manipulations in cloned DNA sequencing and in bioinformatics. The pattern of TIL gene expression was studied in one ovarian cancer and one liver cancer. TILs had three different gene expression profiles: the first set of genes is involved in cell proliferation and mitogenic stimulation, such as c-myc and IL-8, LD78, MIP-1beta, insulin-induced protein and AH-receptor; the second set of genes includes those involved in attachment of lymphocytes to endothelium and extravasation into tumor tissues such as P-selectin ligand and integrin; and the third set, which includes genes such as the perforin, FAS ligand and granzyme B, is related to cytotoxic function to tumor cells. Furthermore the patterns of TIL gene expression obtained from two specimens are marginally different and can be used in explaining the basis of molecular mechanisms regulating cellular interactions and cytotoxicity (Li et al. 2000).

Coulouarn et al. provided a comprehensive dynamic characterization of the cellular and molecular alterations involved in HCC onset and progression in the c-Myc/Tgf-alpha transgenic mouse model. By using a functional genomics approach focused on the early stages of HCC disease, they found that disruption of innate immune surveillance mediated by NK and NKT cells may contribute to oncogenic process in this model of accelerated liver cancer. In particularly, by flow cytometry analysis, they found loss of the major histocompatibility complex class I expression in dysplastic hepatocytes followed by upregulation of numerous activating ligands for NK cells concomitant with a drastic decrease in hepatic NK cell frequency (Coulouarn et al. 2011).

Zhang et al. performed a genomic study combining single-cell mRNA differential display and RNA subtractive hybridization to elucidate CD8 T-cell quiescence/ignorance. By comparing actively maintained quiescent CD8 T cells from liver tumour TILs with quiescent T cells at the single-cell level, they identified differentially expressed candidate genes by high-throughput screening and comparative analysis of expressed sequence tags (ESTs). While genes for the T-cell receptor, tumour necrosis factor (TNF) receptor, TNF-related apoptosis inducing ligand (TRAIL) and perforin were down-regulated, key genes such as Tob, transforming growth factor (TGF)-beta, lung Krupple-like factor (LKLF), Sno-A, Ski, Myc, Ets-2 repressor factor (ERF) and RE1-silencing transcription factor (REST/NRSF) complex were highly expressed in the quiescent TIL CD8 cells. A regulation model is proposed for actively maintained quiescence in CD8 T cells, including three components: up-regulation of the TGF-beta pathway, a shift in the MYC web and inhibition of the cell cycle (Zhang et al. 2009).

9.4.2 Proteomics

In the postgenome era, proteomics and peptidomics are increasingly being used to discover, validate and characterize network biomarkers that can potentially be used for diagnostic purposes and to aid in pharmaceutical development (Goh and Wong 2013). And oncoproteomics is defined as "the study of proteins and their interactions in a cancer cell by proteomic technologies" (Celis and Moreira 2010; Joshi et al. 2011; Cho and Cheng 2007), offering great promise for unveiling the complex molecular events of tumorigenesis (Niu et al. 2010; Mas et al. 2009), as well as those that control clinically important tumor behaviors, such as metastasis, invasion and resistance to therapy (Fig. 9.4). Proteomic is used to identify large-scale changes in regulatory mechanisms involved in the re-programming of T cells during the G0-G1 transition and mapped them to form functionally annotated, dynamic protein interaction networks (Orr et al. 2012). Proteomic study was also used to clarify the mechanism of HCC immune escape concerning lymphocyte dysfunction and

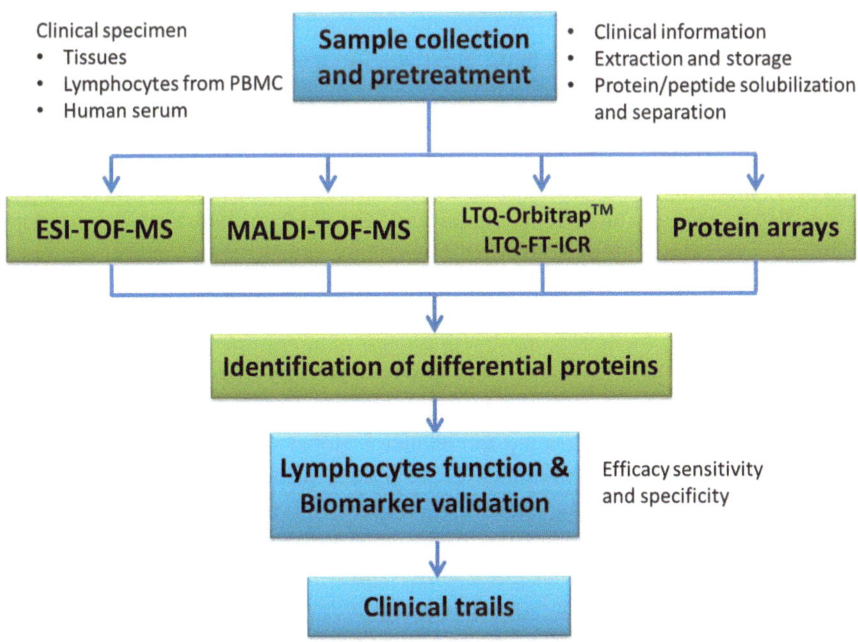

Fig. 9.4 Workflow of proteomic approach

their association with HCC invasion. Weng et al. (2008) analyzed human peripheral blood mononuclear cells (PBMCs) derived DCs pulsed with different metastatic potential human HCC lysates by two-dimensional electrophoresis and electrospray mass spectrometry. The clinical significance of beta-centractin, one of the largest quantitative changed spot, down-regulation in DCs pulsed with high metastatic potential HCC lysates might associate with DCs dysfunction and HCC invasiveness (Weng et al. 2008).

9.4.3 Next-Generation Sequencing (NGS)

The development of cancer is characterized by the joined occurrence of alterations on different levels, from single nucleotide changes via structural and copy number variations to epigenetic alterations (Schweiger et al. 2011). Therefore, analyses of cancer genome sequences and structures provide insights for understanding cancer biology, diagnosis and therapy (Meyerson et al. 2010). Traditional approaches to sequence analysis, including Sanger sequencing, pyrosequencing and allele-specific PCR, are now widely used to guide therapy for cancers. However, traditional sequence analysis has been limited in bandwidth and throughput and as a result, has been focused exclusively on testing the most common aberrations in key genes or fully sequencing single genes (Cronin and Ross 2011).

The application of second-generation DNA sequencing technologies, also known as NGS through whole-genome, whole-exome and whole-transcriptome approaches, has produced a transformation in cancer genomics, generating large data sets that can be analyzed in different ways to answer a multitude of questions about the genomic alterations associated with cancer (Meyerson et al. 2010). For example, pathway-based analyses that provide the full mutational impact on cellular protein networks and correlation analyses aimed at revealing causal relationships between genomic alterations and clinical presentations are both enabled (Ding et al. 2010). NGS can also simultaneously detect deletions, small insertions, copy number alterations, chromosomal rearrangements, translocations and exome-wide base substitutions in all known cancer-related genes (Cronin and Ross 2011; Totoki et al. 2011). NGS platforms have evolved to provide an accurate and comprehensive means for the research on lymphocyte in cancers including HCC (Li and Mao 2013) (Fig. 9.5).

Ding et al. devised a massive anchored parallel sequencing (MAPS) method using next-generation sequencing to conduct a global survey of HBV integration events in 40 pairs of HBV–related HCC tissues (cancer and adjacent tissues). They identified 286 unique integration sites (UISs) with precise HBV–Human DNA junctions. To understand the function of the genes targeted by HBV, they performed Gene Ontology (GO) analysis using the GO miner program, and found that the targeted genes are enriched in several GO terms related to T cell differentiation and activation involved in immune response (Ding et al. 2012).

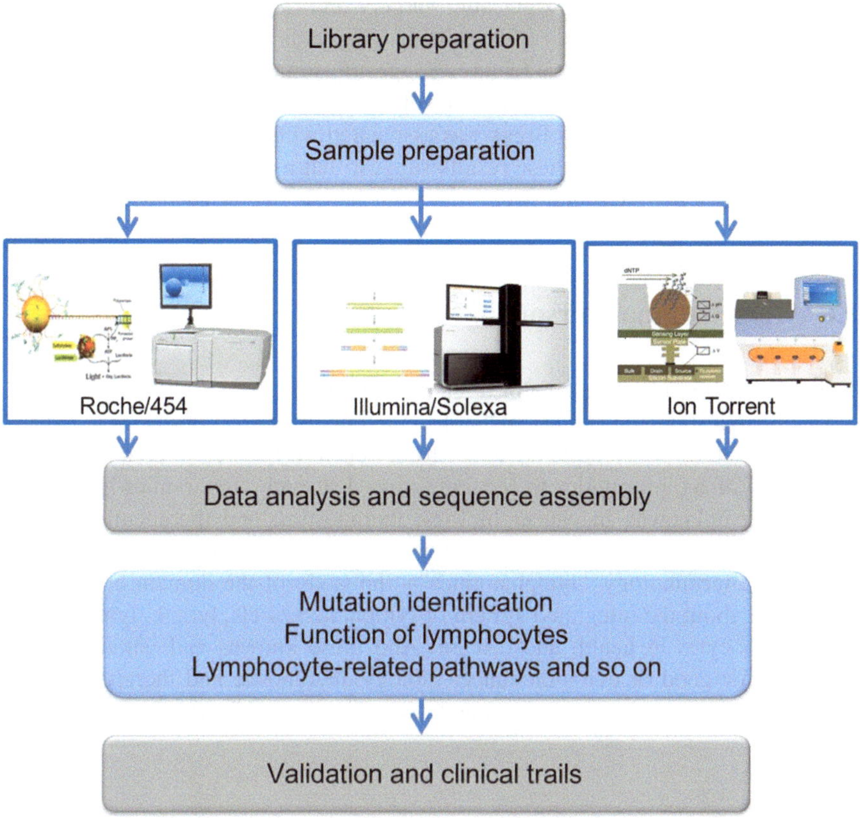

Fig. 9.5 Perspectives of next-generation sequencing in lymphocytes

9.5 Networks in Liver Cancer

Tumor is established as an intersection of malignant clone cells and surrounding stromal cells, which regulate immune responses by coordinating lymphocyte homing, differentiation, activation and antigen response. Hence, researchers emphasized on the intrinsic networks between stromal cells and lymphocytes in the tumor microenvironment in order to find potential immunotherapeutic approaches targeting stroma-lymphocyte interactions (Gorgun and Anderson 2011). Chen et al. (2012) investigated perioperative dynamic alterations of peripheral circulating Tregs and B-regulatory lymphocytes (Bregs) in HCC patients and found that the frequencies of peripheral Tregs and Bregs in HCC patients increased after surgery. The level of circulating CD4(+) CD25(+) CD127(−) Tregs in HCC patients was significantly lower than that in healthy donors and patients with chronic HBV infection before surgery, but was increased after surgery. Preoperative level of CD19(+) IL-10(+) Bregs in HCC patients was also significantly lower than the other groups. However, it was dramatically elevated right after surgery and remained elevated compared to the controls. The frequency of circulating Tregs was correlated with circulating leukocytes, ferritin, and clinical features, suggesting tumor aggressiveness including portal vein thrombosis, hepatic vein involvement and advanced clinical stages. The frequency of circulating Bregs was associated with HBV e antigen and HBV DNA copy number. All of these results suggest that a postoperative combination of therapies against Tregs and Bregs may be beneficial for better outcome of HCC patients after resection.

9.6 Translational/Personalized Medicine

In the elusive quest for "personalized" cancer treatments, diverse challenges must be overcome: questionable validity of "molecular models of life," obstacles to bidirectional translation of scientific advances from bench to bedside to community, and limitations of bioinformatics to recognize and deal with "ignoramics/ignoromes" (expanding unknowns in cancer biology, theranostics, and therapeutic choices) (Witte 2011). These considerations apply in lymphology—or rather in the latest fashionable terminology, "lymphatomics"—the study of the domain or "ome" of everything about the integrated system of lymphatic vessels, lymph, lymph nodes, and lymphocytes in health and disease. And these knowns and unknowns bear directly—for good or evil—on cancer biology, theranostic and therapeutics, and serve as cautionary notes for any concrete blueprint drafted for the future (Fig. 9.6).

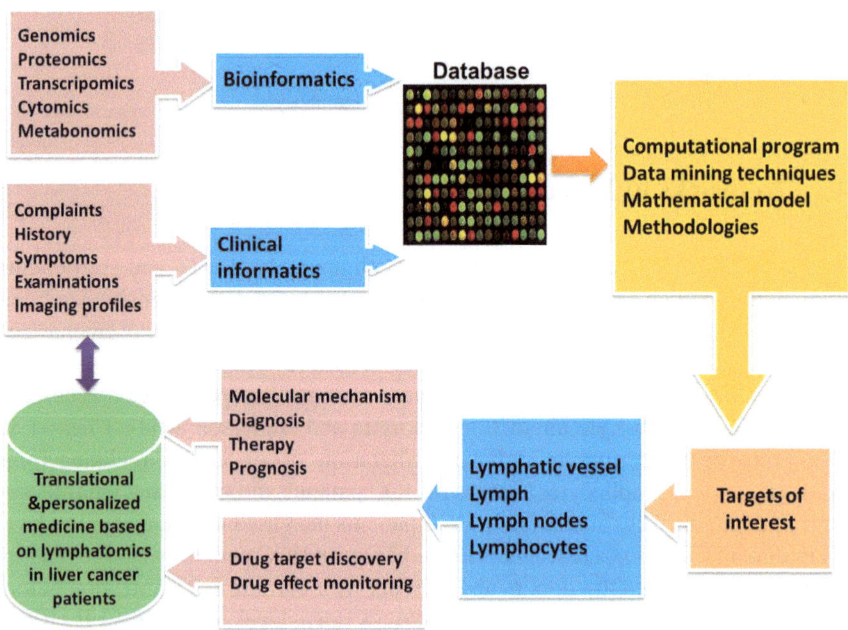

Fig. 9.6 Translational and personalized medicine in "lymphatomics" era

9.7 Conclusion

Both the quantity and quality of TIL are crucial determinants of the outcome of antitumor immune responses. A better understanding of the role of lymphocytes in liver cancer is important for designing future immunotherapy-based clinical protocols. The final goal is to develop accurate tools based on the lymphocyte status of each patient to deliver the right treatment to the right patient at the right time. Clinical bioinformatics can improve the development of lymphocyte targeted therapy in liver cancer. Translational and evidence-based medicine can take advantage of technological advances which offer a variety of high-throughput data for screening the molecular activities of genomic, transcriptional, post-transcriptional and translational studies in the immune system. We envision that lymphocyte-targeted therapy as an effective strategy will become one of the most critical and useful approaches of systematic clinical medicine for liver cancer research and applications that can improve the quality of life of patients with liver cancer.

References

Berzofsky JA, Terabe M. The contrasting roles of NKT cells in tumor immunity. Curr Mol Med. 2009;9(6):667–72.

Buchmann P, Dembek C, Kuklick L, Jager C, Tedjokusumo R, von Freyend MJ, et al. A novel therapeutic hepatitis B vaccine induces cellular and humoral immune responses and breaks tolerance in hepatitis B virus (HBV) transgenic mice. Vaccine. 2013;31(8):1197–203.

Cai XY, Gao Q, Qiu SJ, Ye SL, Wu ZQ, Fan J, et al. Dendritic cell infiltration and prognosis of human hepatocellular carcinoma. J Cancer Res Clin. 2006;132(5):293–301.

Cao MD, Cabrera R, Xu YL, Firpi R, Zhu HZ, Liu C, et al. Hepatocellular carcinoma cell supernatants increase expansion and function of CD4(+)CD25(+) regulatory T cells. Lab Invest. 2007a;87(6):582–90.

Cao DY, Yang JY, Dou KF, Ma LY, Teng ZH. alpha-fetoprotein and interleukin-18 gene-modified dendritic cells effectively stimulate specific type-1 CD4- and CD8-mediated T-cell response from hepatocellular carcinoma patients in vitro. Hum Immunol. 2007b;68(5):334–41.

Celis JE, Moreira JM. Challenges and opportunities in oncoproteomics. Mol Oncol. 2010;4(6):459–60.

Chen TX, Song DL, Min ZH, Wang XD, Gu Y, Wei BJ, et al. Perioperative dynamic alterations in peripheral regulatory T and B cells in patients with hepatocellular carcinoma. J Transl Med. 2012;10:14. doi: 10.1186/1479-5876-10-14.

Cho WC, Cheng CH. Oncoproteomics: current trends and future perspectives. Expert Rev Proteomics. 2007;4(3):401–10.

Coulouarn C, Factor VM, Conner EA, Thorgeirsson SS. Genomic modeling of tumor onset and progression in a mouse model of aggressive human liver cancer. Carcinogenesis. 2011;32(10):1434–40.

Cronin M, Ross JS. Comprehensive next-generation cancer genome sequencing in the era of targeted therapy and personalized oncology. Biomark Med. 2011;5(3):293–305.

Crowe NY, Coquet JM, Berzins SP, Kyparissoudis K, Keating R, Pellicci DG, et al. Differential antitumor immunity mediated by NKT cell subsets in vivo. J Exp Med. 2005;202(9):1279–88.

Ding L, Wendl MC, Koboldt DC, Mardis ER. Analysis of next-generation genomic data in cancer: accomplishments and challenges. Hum Mol Genet. 2010;19(R2):R188–R96.

Ding D, Lou XY, Hua DS, Yu W, Li LS, Wang J, et al. Recurrent targeted genes of hepatitis B virus in the liver cancer genomes identified by a next-generation sequencing-based approach. PLoS Genet. 2012;8:e1003065.

Forner A, Llovet JM, Bruix J. Hepatocellular carcinoma. Lancet. 2012;379(9822):1245–55.
Fu JL, Xu DP, Liu ZW, Shi M, Zhao P, Fu BY, et al. Increased regulatory T cells correlate with CD8 T-cell impairment and poor survival in hepatocellular carcinoma patients. Gastroenterology. 2007;132(7):2328–39.
Gao Q, Qiu SJ, Fan J, Zhou J, Wang XY, Xiao YS, et al. Intratumoral balance of regulatory and cytotoxic T cells is associated with prognosis of hepatocellular carcinoma after resection. J Clin Oncol. 2007;25(18):2586–93.
Gasteiger G, Kastenmuller W. Foxp3+ regulatory T-cells and IL-2: the Moirai of T-cell fates? Front Immunol. 2012;3:179.
Gehring AJ, Xue SA, Ho ZZ, Teoh D, Ruedl C, Chia A, et al. Engineering virus-specific T cells that target HBV infected hepatocytes and hepatocellular carcinoma cell lines. J Hepatol. 2011;55(1):103–10.
Glynne RJ, Ghandour G, Goodnow CC. Genomic-scale gene expression analysis of lymphocyte growth, tolerance and malignancy. Curr Opin Immunol. 2000;12(2):210–14.
Godfrey DI, Uldrich AP, Baxter AG. NKT cells – an early warning system for HBV infection. Nat Med. 2012;18(7):1014–16.
Goh WW, Wong L. Networks in proteomics analysis of cancer. Curr Opin Biotechnol. 2013;24(6):1122–8.
Gorgun G, Anderson KC. Intrinsic modulation of lymphocyte function by stromal cell network: advance in therapeutic targeting of cancer. Immunotherapy. 2011;3(10):1253–64.
Green EA, Gorelik L, McGregor CM, Tran EH, Flavell RA. CD4+CD25+ T regulatory cells control anti-islet CD8+ T cells through TGF-beta-TGF-beta receptor interactions in type 1 diabetes. Proc Natl Acad Sci U S A. 2003;100(19):10878–83.
Hirohashi Y, Sato N. Tumor-associated dendritic cells: molecular mechanisms to suppress antitumor immunity. Immunotherapy. 2011;3(8):945–7.
Hoechst B, Ormandy LA, Ballmaier M, Lehner F, Kruger C, Manns MP, et al. A new population of myeloid-derived suppressor cells in hepatocellular carcinoma patients induces CD4(+)CD25(+)Foxp3(+) T cells. Gastroenterology. 2008;135(1):234–43.
Huang LR, Gabel YA, Graf S, Arzberger S, Kurts C, Heikenwalder M, et al. Transfer of HBV genomes using low doses of adenovirus vectors leads to persistent infection in immune competent mice. Gastroenterology. 2012;142(7):1447–50.
Jemal A, Bray F, Center MM, Ferlay J, Ward E, Forman D. Global cancer statistics. CA Cancer J Clin. 2011;61(2):69–90.
Jonuleit H, Schmitt E, Stassen M, Tuettenberg A, Knop J, Enk AH. Identification and functional characterization of human CD4(+)CD25(+) T cells with regulatory properties isolated from peripheral blood. J Exp Med. 2001;193(11):1285–94.
Joseph RW, Peddareddigari VR, Liu P, Miller PW, Overwijk WW, Bekele NB, et al. Impact of clinical and pathologic features on tumor-infiltrating lymphocyte expansion from surgically excised melanoma metastases for adoptive T-cell therapy. Clin Cancer Res. 2011;17(14):4882–91.
Joshi S, Tiwari AK, Mondal B, Sharma A. Oncoproteomics. Clin Chim Acta. 2011;412(3–4):217–26.
Kato M, Nakamura Y, Suda T, Ozawa Y, Inui N, Seo N, et al. Enhanced anti-tumor immunity by superantigen-pulsed dendritic cells. Cancer Immunol Immunother. 2011;60(7):1029–38.
Kawarabayashi N, Seki S, Hatsuse K, Ohkawa T, Koike Y, Aihara T, et al. Decrease of CD56+ T cells and natural killer cells in cirrhotic livers with hepatitis C may be involved in their susceptibility to hepatocellular carcinoma. Hepatology. 2000;32(5):962–9.
Kirsch IR, Lista F. Lymphocyte-specific genomic instability and risk of lymphoid malignancy. Semin Immunol. 1997;9(3):207–15.
Kobayashi N, Hiraoka N, Yamagami W, Ojima H, Kanai Y, Kosuge T, et al. FOXP3(+) regulatory T cells affect the development and progression of hepatocarcinogenesis. Clin Cancer Res. 2007;13(3):902–11.

Komita H, Homma S, Saotome H, Zeniya M, Ohno T, Toda G. Interferon-gamma produced by interleukin-12-activated tumor infiltrating CD8(+)T cells directly induces apoptosis of mouse hepatocellular carcinoma. J Hepatol. 2006;45(5):662–72.

Kuang DM, Peng C, Zhao QY, Wu Y, Zhu LY, Wang JN, et al. Tumor-activated monocytes promote expansion of IL-17-producing CD8(+) T cells in hepatocellular carcinoma patients. J Immunol. 2010;185(3):1544–9.

Kuiper RP, Schoenmakers EF, van Reijmersdal SV, Hehir-Kwa JY, van Kessel AG, van Leeuwen FN, et al. High-resolution genomic profiling of childhood ALL reveals novel recurrent genetic lesions affecting pathways involved in lymphocyte differentiation and cell cycle progression. Leukemia. 2007;21(6):1258–66.

Lan P, Zhang C, Han Q, Zhang J, Tian Z. Therapeutic recovery of HBV-induced hepatocyte-intrinsic immune defect reverses systemic adaptive immune tolerance. Hepatology. 2013;58(1):73–85.

Lee WC, Wu TJ, Chou HS, Yu MC, Hsu PY, Hsu HY, et al. The impact of CD4(+)CD25(+) T cells in the tumor microenvironment of hepatocellular carcinoma. Surgery. 2012;151(2):213–22.

Li S, Mao M. Next generation sequencing reveals genetic landscape of hepatocellular carcinomas. Cancer Lett. 2013;340(2):247–53.

Li B, Chang T, Larson A, Ding J. Identification of mRNAs expressed in tumor-infiltrating lymphocytes by a strategy for rapid and high throughput screening. Gene. 2000;255(2):273–9.

Liu D, Song L, Wei J, Courtney AN, Gao X, Marinova E, et al. IL-15 protects NKT cells from inhibition by tumor-associated macrophages and enhances antimetastatic activity. J Clin Invest. 2012;122(6):2221–33.

Liyanage UK, Moore TT, Joo HG, Tanaka Y, Herrmann V, Doherty G, et al. Prevalence of regulatory T cells is increased in peripheral blood and tumor microenvironment of patients with pancreas or breast adenocarcinoma. J Immunol. 2002;169(5):2756–61.

Ma Y, Xu YC, Tang L, Zhang Z, Wang J, Wang HX. Cytokine-induced killer (CIK) cell therapy for patients with hepatocellular carcinoma: efficacy and safety. Exp Hematol Oncol. 2012;1(1):11.

Maini MK, Schurich A. The molecular basis of the failed immune response in chronic HBV: therapeutic implications. J Hepatol. 2010;52(4):616–19.

Marchesi V. Immunotherapy: oncolytic vaccinia virus shows promise in liver cancer. Nat Rev Clin Oncol. 2013;10(4):182.

Mas VR, Maluf DG, Archer KJ, Yanek K, Bornstein K, Fisher RA. Proteomic analysis of HCV cirrhosis and HCV-induced HCC: identifying biomarkers for monitoring HCV-cirrhotic patients awaiting liver transplantation. Transplantation. 2009;87(1):143–52.

Mattarollo SR, West AC, Steegh K, Duret H, Paget C, Martin B, et al. NKT cell adjuvant-based tumor vaccine for treatment of myc oncogene-driven mouse B-cell lymphoma. Blood. 2012;120(15):3019–29.

Meyerson M, Gabriel S, Getz G. Advances in understanding cancer genomes through second-generation sequencing. Nat Rev Genet. 2010;11(10):685–96.

Niu D, Feng H, Chen WN. Proteomic analysis of HBV-associated HCC: insights on mechanisms of disease onset and biomarker discovery. J Proteomics. 2010;73(7):1283–90.

Orr SJ, Boutz DR, Wang R, Chronis C, Lea NC, Thayaparan T, et al. Proteomic and protein interaction network analysis of human T lymphocytes during cell-cycle entry. Mol Syst Biol. 2012;8:573.

Patel M, Shariff MI, Ladep NG, Thillainayagam AV, Thomas HC, Khan SA, et al. Hepatocellular carcinoma: diagnostics and screening. J Eval Clin Pract. 2012;18(2):335–42.

Pedroza-Gonzalez A, Verhoef C, Ijzermans JN, Peppelenbosch MP, Kwekkeboom J, Verheij J, et al. Activated tumor-infiltrating CD4+ regulatory T cells restrain antitumor immunity in patients with primary or metastatic liver cancer. Hepatology. 2013;57(1):183–94.

Perz JF, Armstrong GL, Farrington LA, Hutin YJ, Bell BP. The contributions of hepatitis B virus and hepatitis C virus infections to cirrhosis and primary liver cancer worldwide. J Hepatol. 2006;45(4):529–38.

Prieto PA, Durflinger KH, Wunderlich JR, Rosenberg SA, Dudley ME. Enrichment of CD8+ cells from melanoma tumor-infiltrating lymphocyte cultures reveals tumor reactivity for use in adoptive cell therapy. J Immunother. 2010;33(5):547–56.

Radvanyi LG, Bernatchez C, Zhang M, Fox P, Miller P, Chacon J, et al. Specific lymphocyte subsets predict response to adoptive cell therapy using expanded autologous tumor-infiltrating lymphocytes in metastatic melanoma patients. Clin Cancer Res. 2012;18(24):6758–70.

Ren YY, Liu YZ, Ding YP, Song G, Li SH, Wang GQ. Immune characteristics of different immune phases in natural course of chronic HBV infection. Hepatogastroenterology. 2013;60(124):789–95.

Revill P, Yuan Z. New insights into how HBV manipulates the innate immune response to establish acute and persistent infection. Antivir Ther. 2013;18(1):1–15.

Rosenberg SA, Spiess P, Lafreniere R. A new approach to the adoptive immunotherapy of cancer with tumor-infiltrating lymphocytes. Science. 1986;233(4770):1318–21.

Schweiger MR, Kerick M, Timmermann B, Isau M. The power of NGS technologies to delineate the genome organization in cancer: from mutations to structural variations and epigenetic alterations. Cancer Metastasis Rev. 2011;30(2):199–210.

Shalev I, Selzner N, Shyu W, Grant D, Levy G. Role of regulatory T cells in the promotion of transplant tolerance. Liver Transpl. 2012;18(7):761–70.

Shibolet O, Alper R, Zlotogarov L, Thalenfeld B, Engelhardt D, Rabbani E, et al. NKT and CD8 lymphocytes mediate suppression of hepatocellular carcinoma growth via tumor antigen-pulsed dendritic cells. Int J Cancer. 2003;106(2):236–43.

Silva MF, Sherman M. Criteria for liver transplantation for HCC: what should the limits be? J Hepatol. 2011;55(5):1137–47.

Smyth MJ, Godfrey DI. NKT cells and tumor immunity – a double-edged sword. Nat Immunol. 2000;1(6):459–60.

Smyth MJ, Crowe NY, Pellicci DG, Kyparissoudis K, Kelly JM, Takeda K, et al. Sequential production of interferon-gamma by NK1.1(+) T cells and natural killer cells is essential for the antimetastatic effect of alpha-galactosylceramide. Blood. 2002;99(4):1259–66.

Starzl TE. The "privileged" liver and hepatic tolerogenicity. Liver Transpl. 2001;7(10):918–20.

Tanchot C, Terme M, Pere H, Tran T, Benhamouda N, Strioga M, et al. Tumor-infiltrating regulatory T cells: phenotype, role, mechanism of expansion in situ and clinical significance. Cancer Microenviron. 2013;6(2):147–57.

Thimme R, Neagu M, Boettler T, Neumann-Haefelin C, Kersting N, Geissler M, et al. Comprehensive analysis of the alpha-fetoprotein-specific CD8(+) T cell responses in patients with hepatocellular carcinoma. Hepatology. 2008;48(6):1821–33.

Totoki Y, Tatsuno K, Yamamoto S, Arai Y, Hosoda F, Ishikawa S, et al. High-resolution characterization of a hepatocellular carcinoma genome. Nat Genet. 2011;43(5):464–9.

Wada Y, Nakashima O, Kutami R, Yamamoto O, Kojiro M. Clinicopathological study on hepatocellular carcinoma with lymphocytic infiltration. Hepatology. 1998;27(2):407–14.

Wang F, Jing X, Li G, Wang T, Yang B, Zhu Z, et al. Foxp3+ regulatory T cells are associated with the natural history of chronic hepatitis B and poor prognosis of hepatocellular carcinoma. Liver Int. 2012;32(4):644–55.

Weng YQ, Qiu SJ, Liu YK, Fan J, Gao Q, Tang ZY. Down-regulation of beta-centractin might be involved in dendritic cells dysfunction and subsequent hepatocellular carcinoma immune escape: a proteomic study. J Cancer Res Clin Oncol. 2008;134(2):179–86.

Witte MH. Translational/personalized medicine, pharmaco/surgico/radiogenomics, lymphatic spread of cancer, and medical ignoromes. J Surg Oncol. 2011;103(6):501–7.

Zhang W, Ding J, Qu Y, Hu H, Lin M, Datta A, et al. Genomic expression analysis by single-cell mRNA differential display of quiescent CD8 T cells from tumour-infiltrating lymphocytes obtained from in vivo liver tumours. Immunology. 2009;127(1):83–90.

Zhou J, Ding T, Pan WD, Zhu LY, Li L, Zheng LM. Increased intratumoral regulatory T cells are related to intratumoral macrophages and poor prognosis in hepatocellular carcinoma patients. Int J Cancer. 2009;125(7):1640–8.

Xiaojing Xu Department of Medical Oncology, Zhongshan Hospital, Fudan University, Shanghai, China. Her research mainly focuses on the micro-environmental mechanisms of metastasis and recurrence of tumors.

William C.S. Cho RCMP (HK), Ph.D., FHKSMDS, FHKIMLS, FIBMS (UK), Chartered Scientist (UK), Guest Professor. Dr. Cho's main research interests have been focusing on cancer studies utilizing high-throughput technologies to discover biomarkers for cancer diagnosis, treatment prediction and prognostication. Dr. Cho has published over 200 papers and plenty of books covering cancer biomarkers, proteomics, microRNAs and Chinese medicine. He serves as the editor-in-chief, editor and associate editor of a number of international medical journals. He is also an international grant reviewer of the Hope Funds for Cancer Research (USA), Cancer Research (UK), MRC Research Grant (UK), Science Foundation (Ireland), The Foundation Fournier-Majoie for Innovation (Brussels) and National Medical Research Council (Singapore), etc.

Jinglin Xia Project Director. Dr. Xia is the vice president of Medical College, Fudan University, Shanghai, China. He finished postdoctoral research at the Medical Center of Pittsburgh University in 2002–2003. He serves as the editor of a number of international medical journals, executive member of the International Association of translational medicine; vice-chairman of the Shanghai Medical Association. Dr. Xia's main research focuses on tumor angiogenesis and tumor inflammatory microenvironment. Dr. Xia has published over 70 papers and plenty of books covering tumor angiogenesis, microRNAs, and inflammatory factors and so on.

Chapter 10
Early Biomarkers of Hepatocyte Necrosis

Mitchell R. McGill and Hartmut Jaeschke

Abstract Existing serum biomarkers of liver cell death are useful for tracking patient liver injury and recovery. However, few are helpful for early detection or prediction of hepatic damage. There is a definite need for improved non-invasive liver injury biomarkers in the future. Fortunately, progress is being made in that direction. New biomarkers of hepatocyte death, comprising mechanistic biomarkers of apoptosis/necrosis, mitochondrial damage, inflammation, and other pathways, show promise for earlier and more accurate prediction of patient injury and outcome. This chapter provides a brief review of this emerging generation of biomarkers and possibilities for the future.

Keywords Drug-induced liver injury • Biomarker • Necrosis • Apoptosis • Mitochondria • Inflammation

10.1 Introduction

Biomarkers are important in translational research and practice. Broadly defined, a biomarker is a superficial indicator of an underlying physiological or pathophysiological process. In this sense, a biomarker may be a disease symptom (e.g. retinopathy in people with diabetes), an exogenous probe that can be administered to a subject and monitored as it traverses or is altered within the subject's body, or an endogenous molecule that can be detected in serum and which changes in some way in response to a condition of interest. Generally, biomarkers provide rapid and non-invasive means to assess the health of an individual or to gain insight into disease mechanisms.

In the diagnosis and study of liver injury, endogenous serum biomarkers are the most important class. The alanine and aspartate aminotransferases (ALT and AST, respectively) were first proposed as biomarkers of organ injury in the 1950s, when they were shown to increase in serum from patients with myocardial infarction,

M.R. McGill, Ph.D. • H. Jaeschke, Ph.D. (✉)
Department of Pharmacology, Toxicology and Therapeutics, University of Kansas Medical Center, 3901 Rainbow Blvd, MS 1018, Kansas City, KS 66160, USA
e-mail: hjaeschke@kumc.edu

X. Wang (ed.), *Single Cell Sequencing and Systems Immunology*,
Translational Bioinformatics 5, DOI 10.1007/978-94-017-9753-5_10

hepatitis and some forms of cancer (Karmen et al. 1955; LaDue and Wroblewski 1955, 1956). They have since become standard markers of liver injury. Prior to this, time-consuming, inaccurate and imprecise tests based on biotransformation of various compounds or precipitation of serum globulins was used to diagnose liver damage. Although some modern methods to assess liver function, such as serum bilirubin and albumin measurement, were available at the time, in the context of acute liver injury these parameters generally do not change until well after injury has begun – sometimes even days after injury. Aminotransferases are released into circulation at the outset when hepatocytes are dying. Importantly, while both ALT and AST were found to be elevated in serum from liver injury patients in early studies, the discovery that ALT is more abundant in the liver than in other organs led to the use of this enzyme as a specific biomarker of liver damage (LaDue and Wroblewski 1956). As a result, ALT is now widely considered the primary clinical indicator of liver cell injury.

Although measurement of aminotransferases was a major step forward in the clinical setting, these biomarkers are limited by the lack of several important features. For example, aminotransferases are not etiology-specific. In other words, all forms of liver injury cause serum ALT and AST elevation. Thus, aminotransferases are not useful for diagnosis of specific causes of liver injury, such as viral hepatitis or exposure to hepatotoxicants. Serum aminotransferase activities are also not useful for prognosis, as they do not predict patient outcome in liver injury (Antoine et al. 2012; McGill et al. 2014c). In addition, ALT and AST apparently require necrotic cell death for release into circulation, so they cannot be used to detect stress prior to the onset of injury. Finally, aminotransferases do not provide any information about the molecular mechanisms of liver injury.

Due to the shortcomings of ALT, AST and the various liver function tests that are currently used, considerable research is now focused on identifying new biomarkers of liver injury. The purpose of this chapter is to provide a brief overview of the latest developments in this area. The focus is on early biomarkers and those that may be useful for prediction of patient outcomes. It should be noted that acetaminophen (APAP) hepatotoxicity is referred to frequently throughout this chapter. APAP-induced liver injury is a major clinical problem (Larson et al. 2005; Lee 2008) and is relatively easy to study, both clinically and experimentally. Thus, considerable work has been done on biomarkers in APAP hepatotoxicity.

10.2 Biomarkers of Hepatocyte Necrosis and Apoptosis

Two major forms of cell death are widely recognized. Morphologically, apoptosis is characterized by cell shrinkage due to intracellular proteolysis, condensation of nuclear DNA, and membrane blebbing with formation of apoptotic bodies. Throughout the process, the membranes remain intact. On the other hand, cell swelling and cell membrane rupture are typical of oncotic necrosis. While the former is thought of as programmed cell death, the latter is generally considered traumatic; however, both apoptosis and necrosis can involve complex signaling pathways. While apoptosis signaling is now well understood, the prevailing view of necrosis as any spontaneous cell death that is not apoptosis over the previous two decades and the complexity that is now emerging in necrosis pathways has delayed the study of necrosis signaling. Thus, at present, using biomarkers to exclude apoptosis is the best way to demonstrate that a type of liver injury is causing necrosis. One approach to the assessment of apoptosis is to directly measure the activation of apoptosis mediators. There are two general kinds of apoptosis: extrinsic and intrinsic pathways of apoptotic cell death (Fadeel and Orrenius 2005). In the former case, ligation of a death ligand (e.g. tumor necrosis factor (TNF), Fas ligand, etc.) to a death receptor (e.g. TNFR, Fas receptor, etc.) results in activation of caspase 8, and if the signal from the death receptor is strong enough then enough caspase 8 is produced to directly activate caspase 3, which then cleaves cellular proteins (Fig. 10.1). If the death receptor signal is weak, then caspase 8 can only indirectly activate caspase 3 by truncating the B-cell lymphoma 2 (Bcl-2) family member Bid, which then translocates into mitochondria and facilitates assembly of a pore in the outer mitochondrial membrane that permits release of cytochrome c, among other signaling mediators (Fig. 10.1). Cytochrome c combines with apoptotic protease activating factor 1 (Apaf-1) and ATP to form the apoptosome complex, which activates caspase 9. Caspase 9 can then activate caspase 3 (Fig. 10.1). In intrinsic apoptosis, there is sufficient mitochondrial damage to lead to release of cytochrome c and subsequent caspase activation without death receptor signaling. In either case, caspases must be cleaved by other, upstream caspases in order to become active. As the final player in this caspase cascade, caspase 3 activation is a "point of no return" in apoptosis and can thus be measured as a strong indicator of apoptotic signaling. It is possible to measure cleaved caspase 3 in both liver tissue and serum by western blot (McGill et al. 2012) (Fig. 10.1). Alternatively, the protease activity of caspase 3 can be measured directly using caspase 3 substrates which become fluorescent after being cleaved (McGill et al. 2012). Both of these strategies have been shown to work using samples from rats and mice treated with a combination of galactosamine and endotoxin that strongly induces apoptosis in the liver (Gujral et al. 2003; McGill et al. 2012). However, when the same approaches were taken to assess cell death in mice treated with the hepatotoxic drug acetaminophen (APAP), or in APAP overdose patients with liver injury, there was no evidence of caspase 3 activation, suggesting that apoptosis is not a major form of cell death during APAP-induced liver injury in either rodents or humans (Gujral et al. 2002; McGill et al. 2012).

Another strategy to detect apoptosis is to measure the downstream effects of caspase activation. Though less direct, this approach has become much more popular in recent years because the techniques that can be used are thought to be more sensitive. One example is the measurement of caspase-cleaved keratin 18. Keratin 18 is proteolytically cleaved by caspases at an aspartate-alanine-leucine-aspartate/serine (DALD-D/S) amino acid sequence near its C-terminal end (Leers

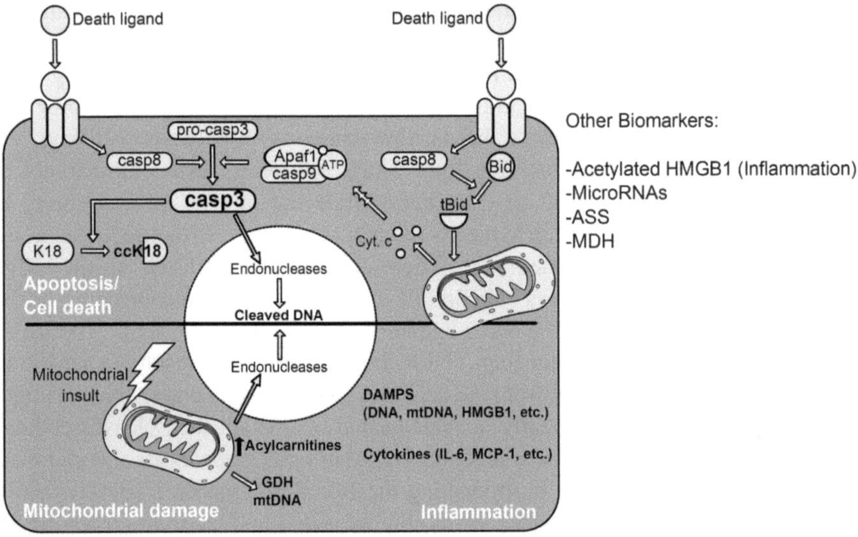

Fig. 10.1 Formation of serum biomarkers within the hepatocyte prior to release into serum. Apoptosis can occur following either direct (Type I) or indirect, mitochondria-mediated (Type II) activation of caspase 3 (casp3). Type II apoptosis involves cleavage of Bid to tBid, which translocates into mitochondria and facilitates release of pro-apoptotic proteins like cytochrome c (Cyt. c). Cyt.c binds to Apaf1 and forms a complex with caspase 9 (casp9) and ATP called the apoptosome. The active casp9 can then cleave and activate casp3. Casp3 cleaves downstream targets, including keratin 18 (K18) to form a caspase-cleaved K18 (ccK18) fragment that can be measured later in serum. Caspases can also activate endonucleases that cleave nuclear DNA. However, this is not specific for apoptotic cell death. Damaged mitochondria release their content into the cytosolic space, including endonucleases, glutamate dehydrogenase (GDH) and mitochondrial DNA (mtDNA). When the cell membrane is compromised, GDH, mtDNA and nuclear DNA fragments can be measured in circulation. Nuclear and mitochondrial DNA can also serve as damage-associated molecular patterns (DAMPs), along with high mobility group box 1 protein (HMGB1) and various other proteins. DAMPs can be measured as possible biomarkers of inflammation, as well as cytokines. Other biomarkers include acetylated HMGB1 (from inflammatory cells), argininosuccinate synthetase (ASS), and malate dehydrogenase (MDH)

et al. 1999), resulting in two smaller peptides (Fig. 10.1). This cleavage exposes an epitope on one of the fragments that can be recognized by an antibody (M30) that is different from the antibody used to detect the full-length K18 (M65) (Leers et al. 1999). It is critical in this approach to measure both the cleaved and full-length forms of K18. The cleaved form of the protein will likely increase in serum even during oncotic necrosis simply because dying cells release their contents, which may include some baseline amount of cleaved K18. Thus, the ratio of M30 signal to M65 signal in the serum is the true measurement of apoptosis. If the ratio is high, as it is in mice treated with galactosamine and endotoxin (Woolbright et al. 2013), then apoptosis is likely occurring. On the other hand, if the ratio is low, then the dominant mode of cell death is probably necrosis (Woolbright et al. 2013; Yang et al. 2014).

Two studies have looked at changes in serum levels of cleaved and full length K18 in APAP-induced liver injury in humans (Antoine et al. 2012, 2013). It was found that the caspase cleaved K18 accounted for only a minor percentage of the total serum K18 (Antoine et al. 2012). Although the serum levels of both cleaved and full length K18 appeared to mirror ALT in most cases, it was noted that there was an increase in both M30 and M65 before ALT in two patients who presented early after APAP overdose (Antoine et al. 2012), suggesting that K18 may be an early biomarker of injury. This was confirmed by the finding that admission levels of full-length K18 are predictive of later ALT increases in drug-induced liver injury (Antoine et al. 2013). It was also reported that patients taking therapeutic doses of APAP who developed elevated ALT levels, indicating liver injury that was otherwise asymptomatic, had increased levels of full-length serum keratin-18 before the increase in ALT (Thulin et al. 2013). Thus, it seems that K18, especially full length K18, may be useful as an early biomarker in acute liver injury and necrosis. Time course experiments have not yet been attempted with direct measurements of caspase-3 activation.

Before moving on, it is worth noting that even currently-used clinical biomarkers of liver damage and cell content release, like ALT and AST, could be considered indicative of cell necrosis. In the apoptosis pathway, cells are disassembled from the inside by proteases, leaving membranes intact, and cell remnants are then cleared by phagocytes. Thus, one would not expect large-scale release of intracellular proteins like ALT. However, secondary necrosis is a phenomenon known to occur in vitro and for which there is also evidence from in vivo studies (Silva 2010). The simple fact that ALT and cytokines are released in large amounts after activation of death receptors in mice (Bajt et al. 2000; Jaeschke and Lemasters 2003) likely demonstrates that secondary necrosis occurs in the liver in animals.

10.3 MicroRNA

MicroRNAs (miRNAs) are short RNAs with only 18–22 nucleotides that help to regulate translation from mRNA to proteins. These small RNA molecules can be released into serum either as a result of cell death or through specific transport processes (Etheridge et al. 2011). MiRNAs have been shown to be elevated in a number of different diseases (Weiland et al. 2012), especially various kinds of liver injury (Bala et al. 2012). In particular, Wang et al. (2009) showed that the serum levels of several miRNAs change before ALT increases in the murine APAP model of acute liver injury. They also demonstrated that these miRNAs increase after treatment with doses of APAP that do not cause ALT elevation. These data suggested that miRNAs may be earlier and more sensitive biomarkers of liver cell death than those currently used. This was confirmed in a study of APAP overdose patients in which it was found that the liver-specific miRNA miR-122 was elevated in serum from humans with APAP-induced liver injury (Starkey-Lewis et al. 2011). Importantly, it was shown that serum miR-122 levels were higher on the first day of study entry in patients with poor outcome than in serum from patients who spontaneously survived . More recently, data were published showing that, as with K18, APAP overdose patients who presented before the development of liver injury had elevated levels of miR-122 that predicted later increases in ALT (Antoine et al. 2013). Additionally, thousands of miRNAs have been identified and serum levels of these small RNA sequences may display unique patterns with different types of tissues injury. In fact, miRNA profiling of human serum identified a panel of miRNAs which are specific for APAP overdose and can discriminate between drug hepatotoxicity and ischemic hepatitis (Ward et al. 2014). Thus, it may be possible to determine the etiology of liver disease in difficult cases simply by measuring the serum miRNA profile. Similarly, existing data support the possibility that miRNA profiles may distinguish between lethal and non-lethal liver injury (Ward et al. 2012). A major advantage of miRNAs as biomarkers is their stability in serum. These RNAs are often associated with extracellular vesicles (exosomes) or secreted proteins that can offer a measure of protection against circulating endonucleases that might degrade them.

10.4 Mitochondrial Damage Biomarkers

Mitochondrial damage is thought to occur in many diseases of the liver, including viral hepatitis (Okuda et al. 2002), APAP-induced liver injury (Meyers et al. 1988; Jaeschke 1990; Kon et al. 2004; Reid et al. 2005; Xie et al. 2014), and alcoholic hepatitis (Thayer 1987; Hirano et al. 1992; Fernandez-Checa and Kaplowitz 2005). However, it is difficult to study mitochondrial dysfunction in humans due to limited availability of tissue samples and because pharmacological interventions cannot be used in patients in the same way as in rodent models. One solution that has been proposed is the development of serum biomarkers of mitochondrial damage (McGill et al. 2012; McGill and Jaeschke 2014). It has been demonstrated that the mitochondrial matrix enzyme glutamate dehydrogenase (GDH), mitochondrial DNA (mtDNA) and acylcarnitines are elevated in serum from mice treated with hepatotoxic doses of APAP and that this only occurs when there is mitochondrial damage (McGill et al. 2012, 2014b) (Fig. 10.1). Nuclear DNA fragments can also be measured in serum after APAP treatment (McGill et al. 2012) and may indirectly indicate mitochondrial damage (Cover et al. 2005). Importantly, elevated levels of GDH, mtDNA and nuclear DNA fragments can also be measured in serum from APAP overdose patients with liver injury (McGill et al. 2012, 2014c). Although only limited time course data are available with these biomarkers, one group has shown that they are higher in the first serum sample drawn after study admission in APAP-induced acute liver failure patients who die compared with those who survive, while ALT levels were not significantly different between the groups (McGill et al. 2014c). These data suggest that these could be earlier biomarkers of liver injury than ALT because they show different kinetics. However, additional research is needed to fully explore that possibility. Moreover, these biomarkers have not yet been measured in samples from patients with other liver diseases.

10.5 Inflammation Biomarkers

A number of studies have been published which seem to show that inflammation is an important mechanism of liver injury (Jaeschke 2011). However, in some cases, e.g. APAP hepatotoxicity, the sterile inflammatory response does not appear to aggravate the injury (Jaeschke et al. 2012). In fact, the main purpose of an inflammatory response to necrotic cell death is to remove the cell debris and prepare for the repair of the damaged tissue. Thus, it is not surprising that biomarkers of liver inflammation would not be elevated early in the time course of injury. Consistent with the latter, circulating neutrophils do not become activated after APAP overdose until after the peak of liver injury in mice or humans (Williams et al. 2010, 2014).

Nevertheless, some inflammatory mediators have been measured in circulation as novel blood biomarkers of liver damage (Fig. 10.1). When cells die, they release proteins and other macromolecules which may act as damage-associated molecular patterns (DAMPs). DAMPs bind to receptors on inflammatory cells and lead to their activation and recruitment. Thus, DAMPs can be considered markers of inflammation, in addition to markers of cell death. One well-studied example of this is the high mobility group box 1 (HMGB1) protein. HMGB1 has a number of functions in both the cytosol and the nucleus, but is more often associated with the nucleus. Severely damaged hepatocytes release HMGB1 into the extracellular space and it can be measured in circulation during liver injury (Antoine et al. 2012; Harrill et al. 2012; Woolbright et al. 2013; Yang et al. 2013b). In addition to release from dying hepatocytes, post-translationally modified forms of HMGB1 can be specifically secreted by inflammatory cells (Bonaldi et al. 2003; Yang et al. 2013a), which adds another layer to the significance of this protein as an inflammation marker. In particular, acetylation controls the subcellular distribution and secretion of HMGB1. Thus, acetylated HMGB1 can be measured in serum as an additional biomarker of inflammation. It has been shown that serum levels of both total and acetylated HMGB1 are associated with worse outcome in APAP hepatotoxicity in humans (Antoine et al. 2012). Unfortunately, while admission levels of total HMGB1 do seem to predict the development of later liver injury (Antoine et al. 2013), it is not yet clear if the acetylated form could serve as an earlier biomarker of liver injury than other parameters. Consistent with the idea that a primary cause of liver injury leads to later inflammation, experiments with the bile duct ligation model (obstructive cholestasis) or hepatic ischemia-reperfusion in mice suggest that acetylated HMGB1 is actually a biomarker of monocyte activation in these models (Woolbright et al. 2013; Yang et al. 2014), and this is consistent with the idea that the late inflammatory response is mainly involved in tissue repair (You et al. 2013). Some cytokines have also been reported to be elevated in APAP overdose patients with liver injury, and these biomarkers may be useful for the differentiation of survivors and non-survivors (James et al. 2001, 2005; Williams et al. 2003). However, detailed time courses have not yet been reported for most of these inflammatory mediators during acute liver injury, especially in humans.

10.6 DNA Damage Biomarkers

As a nuclear protein, HMGB1 could be loosely considered a biomarker of nuclear damage. However, nuclear DNA fragments provide more direct evidence for this mechanism (Fig. 10.1). Nuclear DNA fragments are elevated in serum from patients with acute liver failure from several etiologies, including APAP overdose (Craig et al. 2011; McGill et al. 2012, 2014c), and they correlate with patient outcome in the latter case (McGill et al. 2014c). Limited time course data suggest that nuclear DNA fragments increase and decrease with ALT over time, however, more detailed time course studies are needed to draw stronger conclusions.

10.7 Other Biomarkers

A few other novel biomarkers of liver injury have been described. Malate dehydrogenase (MDH) was found to be elevated in serum from patients with liver injury due to several different etiologies (Schomaker et al. 2013). Unfortunately, only limited time course data were reported, so it is not yet possible to determine if MDH is increased early during hepatic damage. On the other hand, mouse studies revealed that the enzyme argininosuccinate synthetase (ASS) does increase in plasma before ALT, and it was also found to be elevated in plasma from APAP overdose patients (McGill et al. 2014a). ASS was also increased in plasma from mice treated with galactosamine/endotoxin or subjected to hepatic ischemia-reperfusion before an increase in ALT could be detected in either model (Prima et al. 2013). Similar data were reported for sulfotransferase 2a1 (Prima et al. 2013). Future studies should be conducted to determine if circulating ASS and SULT2A1 are predictive of outcome in liver injury patients.

10.8 Conclusions

Better biomarkers of liver cell death are needed. Fortunately, exciting new developments, especially from the study of APAP hepatotoxicity, are moving us closer to this goal. In one particularly interesting recent case report (Dear et al. 2014), a patient who presented to the hospital shortly after an APAP overdose was sent home after blood tests because his serum ALT activity was normal and APAP levels, though high, did not indicate a need for treatment. The patient returned almost 2 days later and was found to have very high ALT activity in serum caused by liver injury. Retrospective measurement of miR-122, K18 and high mobility group box 1 protein (HMGB1) in the patient's serum revealed that all three of these markers were elevated at the time of the patient's first presentation (Dear et al. 2014). These data suggest that, not only are these biomarkers earlier indicators of injury than ALT, but also that they may be used to assess the need for treatment in patients who may be at risk of liver injury but do not yet have any other evidence for it. One major challenge, however, is identifying biomarkers that are both elevated early and that predict patient outcome. In the future, it seems likely that combinations of different serum biomarkers could help to meet this need.

References

Antoine DJ, Jenkins RE, Dear JW, Williams DP, McGill MR, Sharpe MR, Craig DG, Simpson KJ, Jaeschke H, Park BK. Molecular forms of HMGB1 and keratin-18 as mechanistic biomarkers for mode of cell death and prognosis during clinical acetaminophen hepatotoxicity. J Hepatol. 2012;56:1070–9.

Antoine DJ, Dear JW, Lewis PS, Platt V, Coyle J, Masson M, Thanacoody RH, Gray AJ, Webb DJ, Moggs JG, Bateman DN, Goldring CE, Park BK. Mechanistic biomarkers provide early and sensitive detection of acetaminophen-induced acute liver injury at first presentation to hospital. Hepatology. 2013;58:777–87.

Bajt ML, Lawson JA, Vonderfecht SL, Gujral JS, Jaeschke H. Protection against Fas receptor-mediated apoptosis in hepatocytes and nonparenchymal cells by a caspase-8 inhibitor in vivo: evidence for a postmitochondrial processing of caspase-8. Toxicol Sci. 2000;58:109–17.

Bala S, Petrasek J, Mundkur S, Catalano D, Levin I, Ward J, Alao H, Kodys K, Szabo G. Circulating microRNAs in exosomes indicate hepatocyte injury and inflammation in alcoholic, drug-induced, and inflammatory liver diseases. Hepatology. 2012;56:1946–57.

Bonaldi T, Talamo F, Scaffidi P, Ferrera D, Porto A, Bachi A, Rubartelli A, Aqresti A, Bianchi ME. Monocytic cells hyperacetylate chromatin protein HMGB1 to redirect it towards secretion. EMBO J. 2003;22:5551–60.

Cover C, Mansouri A, Knight TR, Bajt ML, Lemasters JJ, Pessayre D, Jaeschke H. Peroxynitrite-induced mitochondrial and endonuclease-mediated nuclear DNA damage in acetaminophen hepatotoxicity. J Pharmacol Exp Ther. 2005;315:879–87.

Craig DG, Lee P, Pryde EA, Masterton GS, Hayes PC, Simpson KJ. Circulating apoptotic and necrotic cell death markers in patients with acute liver injury. Liver Int. 2011;31:1127–36.

Dear JW, Antoine DJ, Starkey-Lewis P, Goldring CE, Park BK. Early detection of paracetamol toxicity using circulating liver microRNA and markers of cell necrosis. Br J Clin Pharmacol. 2014;77:904–5.

Etheridge A, Lee I, Hood L, Galas D, Wang K. Extracellular microRNA: a new source of biomarkers. Mutat Res. 2011;717:85–90.

Fadeel B, Orrenius S. Apoptosis: a basic biological phenomenon with wide-ranging implications in human disease. J Intern Med. 2005;258:479–517.

Fernandez-Checa JC, Kaplowitz N. Hepatic mitochondrial glutathione: transport and role in disease and toxicity. Toxicol Appl Pharmacol. 2005;204:263–73.

Gujral JS, Knight TR, Farhood A, Bajt ML, Jaeschke H. Mode of cell death after acetaminophen overdose in mice: apoptosis or oncotic necrosis? Toxicol Sci. 2002;67:322–8.

Gujral JS, Farhood A, Jaeschke H. Oncotic necrosis and caspase-dependent apoptosis during galactosamine-induced liver injury in rats. Toxicol Appl Pharmacol. 2003;190:37–46.

Harrill AH, Roach J, Fier I, Eaddy JS, Kurtz CL, Antoine DJ, Spencer DM, Kishimoto TK, Pisetksy DS, Park BK, Watkins PB. The effects of heparins on the liver: application of mechanistic serum biomarkers in a randomized study in healthy volunteers. Clin Pharmacol Ther. 2012;92:214–20.

Hirano T, Kaplowitz N, Tsukamoto H, Kamimura S, Fernandez-Checa JC. Hepatic mitochondrial glutathione depletion and progression of experimental alcoholic liver disease in rats. Hepatology. 1992;16:1423–7.

Jaeschke H. Glutathione disulfide formation and oxidant stress during acetaminophen-induced hepatotoxicity in mice in vivo: the protective effect of allopurinol. J Pharmacol Exp Ther. 1990;255:935–41.

Jaeschke H. Reactive oxygen and mechanisms of inflammatory liver injury: present concepts. J Gastroenterol Hepatol. 2011;26 Suppl 1:173–9.

Jaeschke H, Lemasters JJ. Apoptotic versus oncotic necrosis in hepatic ischemia/reperfusion injury. Gastroenterology. 2003;125:1246–57.

Jaeschke H, Williams CD, Ramachandran A, Bajt ML. Acetaminophen hepatotoxicity and repair: the role of sterile inflammation and innate immunity. Liver Int. 2012;32:8–20.

James LP, Farrar HC, Darville TL, Sullivan JE, Givens TG, Kearns GL, Wasserman GS, Simpson PM, Hinson JA. Pediatric Pharmacology Research Unit Network, National Institute of Child Health and Human Development. Clin Pharmacol Ther. 2001;70:280–6.

James LP, Simpson PM, Farrar HC, Kearns GL, Wasserman GS, Blumer JL, Reed MD, Sullivan JE, Hinson JA. Cytokines and toxicity in acetaminophen overdose. J Clin Pharmacol. 2005;45:1165–71.

Karmen A, Wroblewski F, LaDue JS. Transaminase activity in human blood. J Clin Invest. 1955;34:126–31.

Kon K, Kim JS, Jaeschke H, Lemasters JJ. Mitochondrial permeability transition in acetaminophen-induced necrosis and apoptosis of cultured mouse hepatocytes. Hepatology. 2004;40:1170–9.

LaDue JS, Wroblewski F. The significance of the serum glutamic oxalacetic transaminase activity following acute myocardial infarction. Circulation. 1955;11:871–7.

LaDue JS, Wroblewski F. Serum glutamic pyruvic transaminase SGP-T in hepatic disease: a preliminary report. Ann Intern Med. 1956;45:801–11.

Larson AM, Poulson J, Fontana RJ, Davern TJ, Lalani E, Hynan LS, Reisch JS, Schiødt FV, Ostapowicz G, Shakil AO, Lee WM. Acute Liver Failure Study Group. Hepatology. 2005;42:1364–72.

Lee WM. Etiologies of acute liver failure. Semin Liver Dis. 2008;28:142–52.

Leers MP, Kölgen W, Björklund V, Bergman T, Tribbick G, Persson B, Björklund P, Ramaekers FC, Björklund B, Nap M, Jörnvall H, Schutte B. Immunocytochemical detection and mapping of a cytokeratin 18 neo-epitope exposed during early apoptosis. J Pathol. 1999;187:567–72.

McGill MR, Jaeschke H. Mechanistic biomarkers in acetaminophen-induced hepatotoxicity and acute liver failure: from preclinical models to patients. Expert Opin Drug Metab Toxicol. 2014;10:1005–17.

McGill MR, Sharpe MR, Williams CD, Taha M, Curry SC, Jaeschke H. Mechanisms of acetaminophen hepatotoxicity in humans and mice involves mitochondrial damage and nuclear DNA fragmentation. J Clin Invest. 2012;122:1574–83.

McGill MR, Cao M, Svetlov A, Sharpe MR, Williams CD, Curry SC, Farhood A, Jaeschke H, Svetlov SI. Argininosuccinate synthetase as a plasma biomarker of liver injury after acetaminophen overdose in rodents and humans. Biomarkers. 2014a;19:222–30.

McGill MR, Li F, Sharpe MR, Williams CD, Curry SC, Ma X, Jaeschke H. Circulating acylcarnitines as biomarkers of mitochondrial dysfunction after acetaminophen overdose in mice and humans. Arch Toxicol. 2014b;88:391–401.

McGill MR, Staggs VS, Sharpe MR, Lee WM, Jaeschke H, Acute Liver Failure Study Group. Serum mitochondrial biomarkers and damage-associated molecular patterns are higher in acetaminophen overdose patients with poor outcome. Hepatology. 2014c;60:1336–45.

Meyers LL, Beierschmitt WP, Khairallah EA, Cohen SD. Acetaminophen-induced inhibition of hepatic mitochondrial respiration in mice. Toxicol Appl Pharmacol. 1988;93:378–87.

Okuda M, Li K, Beard MR, Showalter LA, Scholle F, Lemon SM, Weinman SA. Mitochondrial injury, oxidative stress, and antioxidant gene expression are induced by hepatitis C virus core protein. Gastroenterology. 2002;122:366–75.

Prima V, Cao M, Svetlov SI. ASS and SULT2A1 are novel and sensitive biomarkers of acute hepatic injury – a comparative study in animal models. J Liver. 2013;2:1.

Reid AB, Kurten RC, McCullough SS, Brock RW, Hinson JA. Mechanisms of acetaminophen-induced hepatotoxicity: role of oxidative stress and mitochondrial permeability transition in freshly isolated mouse hepatocytes. J Pharmacol Exp Ther. 2005;312:509–16.

Schomaker S, Warner R, Bock J, Johnson K, Potter D, Van Winkle J, Aubrecht J. Assessment of emerging biomarkers of liver injury in human subjects. Toxicol Sci. 2013;132:276–83.

Silva MT. Secondary necrosis: the natural outcome of the complete apoptotic program. FEBS Lett. 2010;584:4491–9.

Starkey-Lewis PJ, Dear J, Platt V, Simpson KJ, Craig DG, Antoine DJ, French NS, Dhaun N, Webb DJ, Costello EM, Neoptolemos JP, Moggs J, Goldring CE, Park BK. Circulating microRNAs as potential markers of human drug-induced liver injury. Hepatology. 2011;54:1767–76.

Thayer WS. Effects of ethanol on proteins of mitochondrial membranes. Ann N Y Acad Sci. 1987;492:193–206.

Thulin P, Nordahl G, Gry M, Yimer G, Aklillu E, Makonnen E, Aderaye G, Lindquist L, Mattsson CM, Ekblom B, Antoine DJ, Park BK, Linder S, Harrill AH, Watkins PB, Glinghammar B, Schuppe-Koistinen I. Keratin-18 and microRNA-122 complement alanine aminotransferase as novel safety biomarkers for drug-induced liver injury in two human cohorts. Liver Int. 2013;34(3):367–78.

Wang K, Zhang S, Marzolf B, Troisch P, Brightman A, Hu Z, Hood LE, Galas DJ. Circulating microRNAs, potential biomarkers for drug-induced liver injury. Proc Natl Acad Sci U S A. 2009;106:4402–7.

Ward J, Bala S, Petrasek J, Szabo G. Plasma microRNA profiles distinguish lethal injury in acetaminophen toxicity: a research study. World J Gastroenterol. 2012;18:2798–804.

Ward J, Kanchagar C, Veksler-Lublinsky I, Lee R, McGill MR, Jaeschke H, Curry SC, Ambros V. Circulating microRNA profiles in human patients with acetaminophen hepatotoxicity and ischemic hepatitis. Proc Natl Acad Sci U S A. 2014;111:12169–74.

Weiland M, Gao XH, Zhou L, Mi QS. Small RNAs have a large impact: circulating microRNAs as biomarkers for human diseases. RNA Biol. 2012;9:850–9.

Williams AM, Langley PG, Osei-Hwediah J, Wendon JA, Hughes RD. Hyaluronic acid and endothelial damage due to paracetamol-induced hepatotoxicity. Liver Int. 2003;23:110–15.

Williams CD, Bajt ML, Farhood A, Jaeschke H. Acetaminophen-induced hepatic neutrophil accumulation and inflammatory liver injury in CD18-deficient mice. Liver Int. 2010;30:1280–92.

Williams CD, Bajt ML, Sharpe MR, McGill MR, Farhood A, Jaeschke H. Neutrophil activation during acetaminophen hepatotoxicity and repair in mice and humans. Toxicol Appl Pharmacol. 2014;275:122–33.

Woolbright BL, Antoine DJ, Jenkins RE, Bajt ML, Park BK, Jaeschke H. Plasma biomarkers of liver injury and inflammation demonstrate a lack of apoptosis during obstructive cholestasis in mice. Toxicol Appl Pharmacol. 2013;273:524–31.

Xie Y, McGill MR, Dorko K, Kumer SC, Schmitt TM, Forster J, Jaeschke H. Mechanisms of acetaminophen-induced cell death in primary human hepatocytes. Toxicol Appl Pharmacol. 2014;279:266–74.

Yang H, Antoine DJ, Andersson U, Tracey KJ. The many faces of HMGB1: molecular structure-functional activity in inflammation, apoptosis, and chemotaxis. J Leukoc Biol. 2013a;93:865–73.

Yang Z, Deng Y, Su D, Tian J, Gao Y, He Z, Wang X. TLR4 as a receptor for HMGB1-mediated acute lung injury after liver ischemia/reperfusion injury. Lab Invest. 2013b;93:792–800.

Yang M, Antoine DJ, Weemhoff JL, Jenkins RE, Park BK, Jaeschke H. Biomarkers distinguish apoptotic and necrotic cell death during hepatic ischemia-reperfusion injury in mice. Liver Transpl. 2014;20(11):1372–82.

You Q, Holt M, Yin H, Li G, Hu CJ, Ju C. Role of hepatic resident and infiltrating macrophages in liver repair after acute injury. Biochem Pharmacol. 2013;86:836–43.

Mitchell R. McGill is a postdoctoral fellow at the University of Kansas Medical Center. His primary research interest is acetaminophen hepatotoxicity, with particular emphasis on the development and application of mechanistic biomarkers to study the pathogenesis of acetaminophen-induced acute liver injury in humans. As part of a team of basic scientists and clinicians, he has provided the first evidence that mitochondrial damage occurs in humans after acetaminophen overdose, similar to mice, and that this damage is critical in the mechanisms of toxicity. In addition to this translational work, he also studies the basic pathophysiology of drug-induced liver injury using both in vivo animal models and in vitro human systems. He has authored or co-authored more than 30 journal publications on drug hepatotoxicity, as well as several book chapters. He also serves as a reviewer for multiple journals in the toxicology and hepatology fields.

Hartmut Jaeschke is Professor and Chairman of the Department of Pharmacology, Toxicology, and Therapeutics at the University of Kansas Medical Center, Kansas, USA. He joined KUMC in 2006. He is a Fellow of the Academy of Toxicological Sciences (ATS). After receiving his Ph.D. degree in toxicology from the University of Tübingen, Germany, in 1983, he did his postdoctoral work at the University of Tübingen and at Baylor College of Medicine in Houston, Texas, USA. Since 1987, he held faculty positions at Baylor (1987–1992), at the University of Arkansas for Medical Sciences, Little Rock, (1999–2002), and at the University of Arizona, Tucson (2002–2006). In addition, he spent 6 years in the pharmaceutical industry (The Upjohn Co. and Pharmacia & Upjohn Co.), from 1992 to 1998. He has published more than 300 peer-reviewed papers, invited reviews, and book chapters in the areas of liver toxicology and liver pathophysiology. In the past, he served on the Editorial Boards of *American Journal of Physiology: Liver and Gastrointestinal Physiology, Gastroenterology, and Toxicology Letters* and was an Associate Editor for *Hepatology* (2002–2006) and *Liver International* (2007–2012). He currently serves as Associate Editor of *Toxicological Sciences, BioMed Research International, Inflammation & Cell Signaling, and MO Journal of Toxicology.* In addition, he is a Member of 12 Editorial Boards including *Hepatology*, *Journal of Hepatology*, and *World Journal of Gastroenterology*. He is a member of many national and international scientific societies including the Society of Toxicology and the American Association for the Study of Liver Diseases and served as a member of the Hepatobiliary Pathophysiology Study Section of the National Institutes of Health (2008–2013) and on many NIH review committees for Center Grants or Program Project Grants. Dr. Jaeschke's major research interests include basic mechanisms and translational aspects of xenobiotic-induced hepatotoxicity and tissue repair with focus on acetaminophen overdose, mechanisms of inflammatory liver injury in models of hepatic ischemia-reperfusion injury, obstructive cholestasis, endotoxin shock, signaling pathways of apoptosis and programmed necrosis in hepatocytes.

Chapter 11
Protocol for Single Cell Isolation by Flow Cytometry

Jian Wang, Zhihui Min, Meiling Jin, and Xiangdong Wang

Abstract Single-cell analysis is increasingly applied in biological and clinical studies with the improvement of single-cell methods. Here, we introduce a protocol for single cell isolation by flow cytometry. The main processes include tissue collection, preparation of single-cell suspensions, staining cells with CFSE, fluorescence-activated cell sorting (FACS) for single-cell sorting, and observing and taking photos of the sing cell. The protocol is suitable for tumor tissues but can also be optional for other specimens. It is critical requirement to keep the viability of tissues and cells at low temperature and choose a suitable conditional medium.

Keywords Single cell isolation • Flow cytometry • Fluorescence-activated cell sorting • Staining cells

11.1 Background

It was always assumed that all experimental measurements from cells either in vitro or in vivo represented the property and function of the cell. However, each cell is different, even from the same cell type or the same tissue. The heterogeneity of

---

J. Wang
Shanghai Clinical Bioinformatics Research Institute, Shanghai, China

Z. Min, M.S.
Center for Biomedical Research, Zhongshan Hospital, Fudan University Medical School, Shanghai, China

Fudan University Center for Clinical Bioinformatics, Shanghai, China

M. Jin, M.D.
Department of Pulmonary Medicine, Zhongshan Hospital, Fudan University Medical School, Shanghai, China

X. Wang, M.D., Ph.D. (✉)
Zhongshan Hospital, Fudan University, China

Shanghai Institute of Clinical Bioinformatics, China
e-mail: xiangdong.wang@clintransmed.org

X. Wang (ed.), *Single Cell Sequencing and Systems Immunology*,
Translational Bioinformatics 5, DOI 10.1007/978-94-017-9753-5_11

cells should be considered in the research for precise and deep study (Shapiro et al. 2013). Now the alternative to solve the problem is to study and measure the cell in the single-cell level. A single cell, which weighs a few ng with a size of ~10 um, is the smallest unit of life that can replicate independently and play biological function (Lindstrom and Andersson-Svahn 2010; Hausman 2007). It consists of protoplasm enclosed within a membrane and contains many different biomolecules and inorganic ions. Proteins and nucleic acids as the critical biomolecules only make up less than 25 % weight of the cell. It has been estimated that the molar concentration of a gene is only about 10^{-12} M while the total protein is up to 10^{-9} molecules per cell (hundreds of pg) (Hausman 2007).

Based on the understanding, it is really a big challenge for researchers to perform the DNA sequence and molecular analysis in a single-cell level without approaches and devices improvement. The single-cell sorting methods and the next-generation sequencing approaches were consecutively presented in the last decades, which developed a new research road for the routine experiments (Navin et al. 2011). The next generation sequencing approaches have been improved in precision and throughput with less amounts of DNA for analysis, such as entire genomes from a single cell (Geigl and Speicher 2007). Also, technologies for single cell sorting, such as flow cytometry and microfluidics, make it feasible to isolate single cell from tissue and analyze DNA content in single-cell level (Lindstrom and Andersson-Svahn 2010; Swennenhuis et al. 2013).

Single cell sequencing approaches have been widely applied in different fields. For example, the study of microorganisms that cannot be cultured by direct single-cell genome sequencing (Kalisky et al. 2011), analysis of rare and circulating tumour cells to evaluate the prognosis (Ramskold et al. 2012), the study of tumour heterogeneity and microevolution (Dalerba et al. 2011; Cristofanilli et al. 2005), characterization of the earliest differentiation events in human embryogenesis, and the investigation of transcriptional noise and stochastic fate choice. The present chapter is to share our knowledge and experience of the single cell isolation and introduce the practical protocol of FACS-based single cell isolation.

11.2 Procedure

11.2.1 Tissue Collection

Tumor tissue samples are obtained from surgery extraction of patients and kept in PBS in sterile 50 mL tubes and transported to lab on cold ice as soon as possible. The pathological nature of tissue samples are subsequently confirmed by pathology. It is important to transport tissue samples in cold to keep cells viable.

11.2.2 Preparation of Single-Cell Suspensions

All subsequent procedures should be performed in the tissue-culture flow hood with ultraviolet irradiation in advance to ensure sterile condition. All instruments including forceps and scissors must be received high temperature and high pressure sterilization in advance.

1. Place the tissue sample in a sterile 60 mm Petri dish and add a little normal saline (NS) to keep the tissue wet.
2. Cut off the necrotic areas, fatty tissue, blood clots, and connective tissue with forceps and scissors and rinse the tissue samples with NS.
3. Cut the tumor tissue into very small pieces (approximately 1 mm^2) with scissors and transfer it into 50 ml sterile centrifuge tube.
4. Add the collagenase solution (1 mg/L) 30 ml and incubate for 30 min at 37 °C with gentle stirring.
5. Filter the digested tissues through 100 um sieve into 50 ml sterile centrifuge tube to remove the undigested fibrous tissue.
6. Centrifuge tumor cell suspension at 1,000 rpm for 10 min, discard the supernatant and resuspend in DMEM complete medium. Then repeat this process once more.
7. Count the number with cell counting chamber under microscope and adjust the cell density to $10^6 \sim 10^7$/ml. Store cell suspension on ice for next process.

11.2.3 Staining Cells with CFSE

The staining process must be performed in the sterile tissue-culture flow hood. Carboxyfluorescein diacetate succinimidyl diester (CFSE, Sigma) is fluorescent molecule that can unbiased label the viable cells and has been applied for cell proliferation assay either in vitro or in vivo.

1. Centrifuge the suspension at 800 rpm for 5 min to remove cell fragment and resuspend with 1 ml PBS with 10 %FBS.
2. Add CFSE to 1 uM final concentration and incubate in dark place for 8 min at 37 °C
3. Add the same amount of complete medium with 10 %FBS to terminate the reaction.
4. Centrifuge the suspension at 1,000 rpm for 5 min, discard the supernatant and resuspend in complete medium with 10 %FBS. Repeat this process three times and resuspend with 1 ml PBS at last time.
5. Roughly evaluate the staining cells under fluorescence microscope with blue light and store cells on ice from light for next process.

Fig. 11.1 FACSAriaIII (BD) is used for single cell isolation

11.2.4 Fluorescence-Activated Cell Sorting (FACS) for Single Cell Sorting

FACSAriaIII (BD) is used to sort single cell into 96 well plates (Fig. 11.1). The room must be kept clean and irradiated with ultraviolet light overnight in advance. Prepare conditioned medium and add medium into 96 well plates before single cell sorting. Fill each well with 30 ml in 96 well plates at first. The conditioned medium is made up of the DMEM medium supplemented with 10 % FBS, insulin (5 U/ml), EGF (10 ng/ml), penicillin G (100 U/ml) and streptomycin (100 mg/ml).

1. Turn on and prepare the cell sorter – FACSAriaIII (BD) and open the Diva software on computer
2. Install 100 μm nozzle and select corresponding nozzle size in Sort Setup. The nozzle should be washed by ultrasound for 1 min in advance
3. Calibrate the cell sorter and adjust frequency and amplitude of drop to make the flow break off point in the 1/3~2/3 of the stream window and drop gap value remain stable. Load CS&T quality control microspheres under software indication and machine complete quality inspection automatically.
4. Regulate declination stream and active test sort window to present four straws of declination flow. Merge far left and left flow into one flow and close far right and right flow. Then adjust drop1 and gap value to make drop1 stable.
5. Regulate drop delay with accudrop beads. Open auto delay software to set the drop delay value automatically.

6. Put 96 well plates with lid on ACDU and move to default position by click “Go to Home” icon. Then open test drop and make the drop fall to the lid of 96 well plates. Fine adjustment is needed to make the drop locate in the center of the well. Then load the prepared cells and repeat above process to ensure cells to fall to the center of the well exactly. Open the ACDU refrigeration cycle and set temperature at 4 °C before this step, which is critical to keep cells viable.
7. Set scatter diagram (FSC vs. SSC) and draw P1 gate to select cells. Then set P2 gate to sort CFSE-positive cells. Take off the lid and sort single cell into 96 well plates.
8. Observe the single cell in each well by microscope quickly and put into cell culture incubator with 5 % CO_2 at 37 °C.

11.2.5 Observing and Taking Photos of the Single Cell

Observe the single cell in 96 well plates by microscope at 24 h, 3 day and 7 day after sorting (Fig. 11.2). Each observation time is advised to be limited into 1 h to avoid damaging the viablity of the single cell. Draw a cross on the lid of the well which has no single cell to be found. Take photos of each well with single cells (Figs. 11.3 and 11.4). The medium is advised to be changed every 3 day or longer.

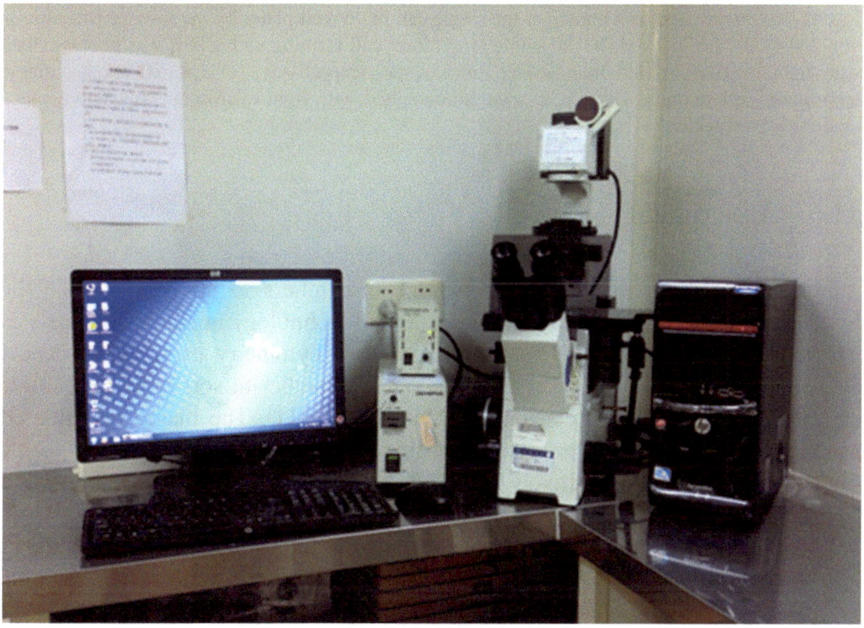

Fig. 11.2 Inversed Fluorescent Microscope for single cell observation

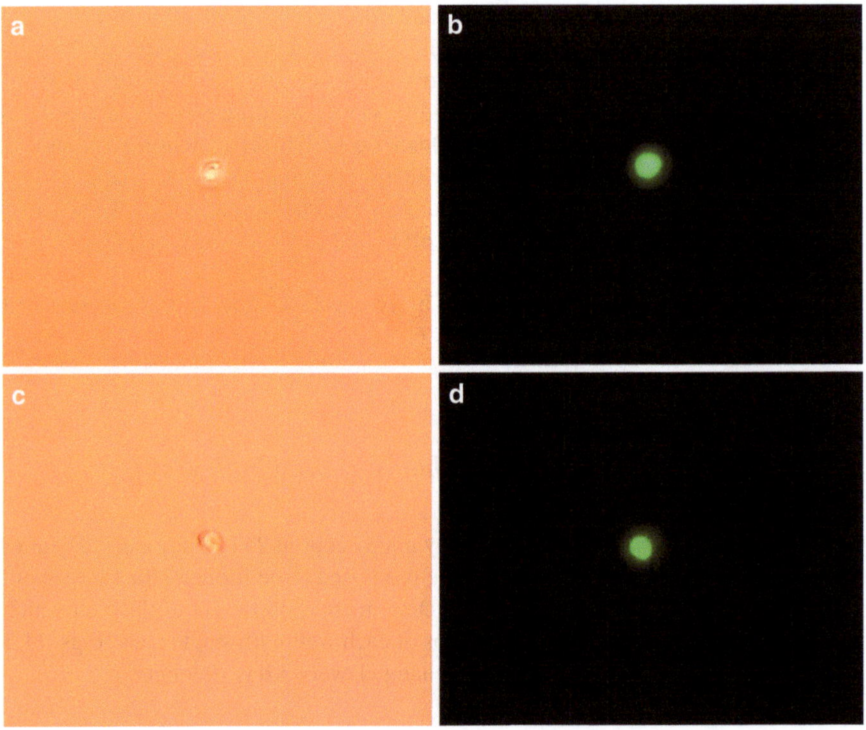

Fig. 11.3 Observe and take photos of the single cell in 96 well plates by microscope (magnification 200×) at 24 h. (**a**) and (**b**) the single lung tumor cell staining with CFSE was photographed with light microscope and fluorescence microscope, respectively. (**c**) and (**d**) another single lung tumor cell staining with CFSE was photographed with light microscope and fluorescence microscope, respectively

11.3 Assessment

Single cell analysis makes it possible for researchers to study and trace the heterogeneity of cells. Flow cytometry as a vital technology in single cell analysis has been widely applied. The advantage of flow cytometry using fluorescence-activated cell sorting (FACS) is that it allows researcher to sort individual single cells with accuracy and precision (Dalerba et al. 2011). The biased or unbiased sample can be acquired by using cell-specific fluorescent antibody or non-specific fluorescent molecule, respectively. We also succeed to unbiased single cell isolation by using light-scattering properties of cells. Furthermore, high-throughput single cell isolation, time-saving property and automation is also the main advantage of FACS-based single cell isolation.

However, the cells as sorting material should be in a large mount ($5\times10^5 \sim 1\times 10^6$/ml), which has a great limit to low-abundance cell subpopulations. The rapid fluid in the machine and non-specific fluorescent molecule can damage the viability

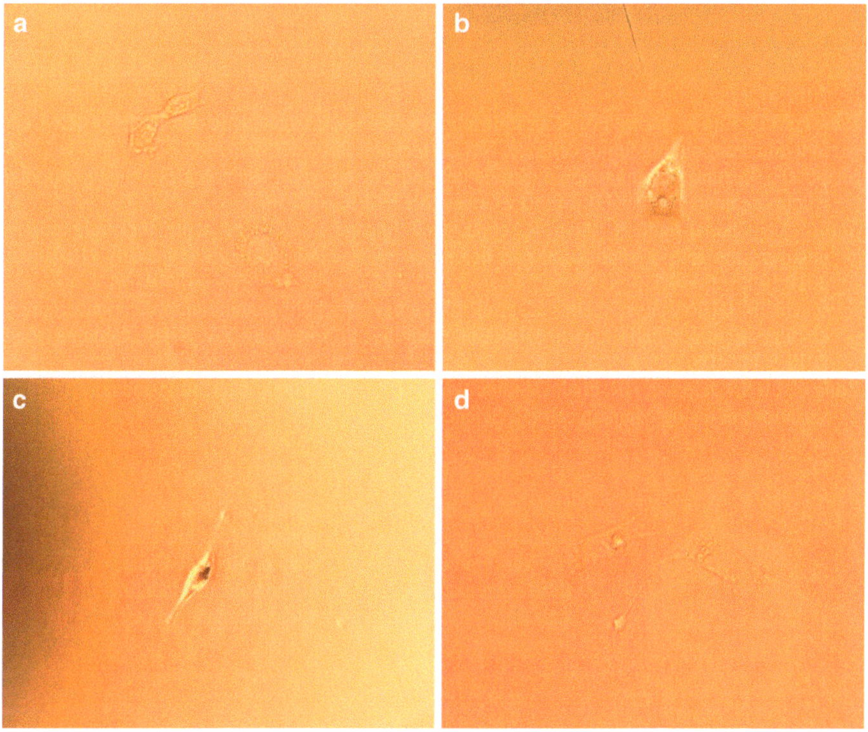

Fig. 11.4 Observe and take photos of the single cell in 96 well plates by microscope on day 3 (Magnification ×200)

of the sorted cells and make the isolation failure. FACS-based single cell isolation also has higher cost compared with other single cell isolation technologies, such as serial dilatation (Shlush et al. 2012) and micromanipulation by using a simple mouth pipette (Zong et al. 2012).

References

Cristofanilli M, Hayes DF, Budd GT, et al. Circulating tumor cells: a novel prognostic factor for newly diagnosed metastatic breast cancer [J]. J Clin Oncol. 2005;23(7):1420–30.

Dalerba P, Kalisky T, Sahoo D, et al. Single-cell dissection of transcriptional heterogeneity in human colon tumors [J]. Nat Biotechnol. 2011;29(12):1120–7.

Geigl JB, Speicher MR. Single-cell isolation from cell suspensions and whole genome amplification from single cells to provide templates for CGH analysis [J]. Nat Protoc. 2007;2(12):3173–84.

Hausman GMCR. The cell [M]. 4th ed. Washington, DC: ASM Press; 2007.

Kalisky T, Blainey P, Quake SR. Genomic analysis at the single-cell level [J]. Annu Rev Genet. 2011;45:431–45.

Lindstrom S, Andersson-Svahn H. Overview of single-cell analyses: microdevices and applications [J]. Lab Chip. 2010;10(24):3363–72.
Navin N, Kendall J, Troge J, et al. Tumour evolution inferred by single-cell sequencing [J]. Nature. 2011;472(7341):90–4.
Ramskold D, Luo S, Wang YC, et al. Full-length mRNA-seq from single-cell levels of RNA and individual circulating tumor cells [J]. Nat Biotechnol. 2012;30(8):777–82.
Shapiro E, Biezuner T, Linnarsson S. Single-cell sequencing-based technologies will revolutionize whole-organism science [J]. Nat Rev Genet. 2013;14(9):618–30.
Shlush LI, Chapal-Ilani N, Adar R, et al. Cell lineage analysis of acute leukemia relapse uncovers the role of replication-rate heterogeneity and microsatellite instability [J]. Blood. 2012;120(3):603–12.
Swennenhuis JF, Reumers J, Thys K, et al. Efficiency of whole genome amplification of single circulating tumor cells enriched by Cell Search and sorted by FACS [J]. Genome Med. 2013;5(11):106.
Zong C, Lu S, Chapman AR, et al. Genome-wide detection of single-nucleotide and copy-number variations of a single human cell [J]. Science. 2012;338(6114):1622–6.

Jian Wang now is studying for a medical master's degree on pulmonary medicine in Zhongshan Hospital, Fudan University. He was graduated from Wenzhou Medical College and had a short-term exchange as visiting student in Medizinische Universität Graz. His work mainly focused on single cell isolation and sequencing.

Zhihui Min MS, is an engineer of Biomedical Research Center of Zhongshan Hospital, Fudan University. Her main research is focused on molecular immunity for tumor and application for Flow Cytometry. She got her master degree in Biochemistry and Molecular Biology for medical college of Lanzhou University. After graduation she had been worked at technology platform of key laboratory of nutrition and metabolism for 6 years in institute of nutritional science, Shanghai institute for biological sciences, Chinese Academy of Sciences. She was responsible for management and operation the instrument for Flow Cytometry, Dual Energy X-ray Absorptiometry and Confocal, assisted in the team to complete the instrument related experiments.

Meiling Jin M.D., chief physician in department of pulmonary medicine, Zhongshan Hospital, Fudan University. She engaged in clinical medicine and research for more than 20 years and specialized in diagnosis and treatment for asthma, lung cancer and COPD. She took charge of multiple research programs and has finished cooperation program with NIH "Dietary Antioxidants and Incidence of Adult-onset Asthma". She has published more than 60 papers, including 6 SCI papers.

Xiangdong Wang Dr. Wang works as a distinguished professor of Respiratory Medicine at Fudan University, China, adjunct professor of Molecular Bioscience at North Carolina State University, USA, and adjunct professor of Clinical Bioinformatics at Lund University. He serves as a Director of Biomedical Research Center, Fudan University Zhongshan Hospital. He was also pointed as visiting, guest and honor professors in Harbin Medical University (China), Zhejiang University (China), Kyoto Prefectural University of Medicine (Japan), and National Academy of Medicine (Romania). Dr. Wang acts as an Editor-in-Chief of Clinical & Translational Medicine, Journal of Clinical Bioinformatics and Journal of Epithelial Biology & Pharmacology, and Asian Editor of Journal of Cellular & Molecular Medicine. His main research is focused on the role of epithelial cells in the development of organ dysfunction, disease-specific biomarkers, and potential therapies.

Chapter 12
A Simple Protocol for Single Lung Cancer Cell Isolation-Making the Single Cell Based Lung Cancer Research Feasible for Individual Investigator

Ding Zhang and Xiangdong Wang

Abstract There are no two identical cells in any organism and with human more so. Quantitative measurements of functions in single cell is a critical challenge in any modern biological study. Now with the development of single cell isolation technology, the era of single cell based research is coming, especially when combination with omics science, which takes full advantage of both techniques, and gets rid of their disadvantages. There are several methods for single cell isolation: serial dilution, micromanipulation (mechanical-using a simple mouth pipette and optimal), fluorescence-activated cell sorting (FACS), laser-capture microdissection (LCM) and microfluidic devices. But some techniques need very expensive equipments, which is not easily available for most investigators. Here we developed a simple protocol to isolate and culture single lung cancer cell. This will make single cell-based lung cancer research much more available for individual investigator.

Keywords Single-cell • Omics • Human • Disease • Preparation

Tissues typically consist of distinct cell types with different ratio, leading to the invariable heterogeneity of human body. All cells in an individual organism were initially thought to have identical genomes (genomics), which turns out to be completely wrong. Somatic mutation is very common during normal cell division and continuously accumulated since the zygotic stage to endow each cell in the body with a unique genomic signature (Frumkin et al. 2005). Even though the genomic signatures are exactly identical, they have different transcriptomes

D. Zhang (✉)
Shanghai Institute of Clinical Bioinformatics, Shanghai, China
e-mail: dingzhang0814@sina.com

X. Wang, M.D., Ph.D. (✉)
Zhongshan Hospital, Fudan University, China

Shanghai Institute of Clinical Bioinformatics, China
e-mail: xiangdong.wang@clintransmed.org

X. Wang (ed.), *Single Cell Sequencing and Systems Immunology*,
Translational Bioinformatics 5, DOI 10.1007/978-94-017-9753-5_12

(transcriptomics), which is determined by their epigenetic state (epigenomics), and have diverse functional proteins (proteomics), metabolic state (metabolomics) and lipidomics. It is impossible to determine whether two cells are actually different in general experiments, while advanced methodologies enable quantitative measurements of functions in small numbers of cells, ideally single cells, represented a critical challenge in any modern biological study.

Omics science is featured by focusing on integrative systems biology. The advent of high-throughput omics technologies (e.g. genomics, transcriptomics, proteomics, metabolomics, lipidomics etc.) has led to a revolution in human diseases and medicine, which offers the promise to investigate the much more complex, longitudinal, and dynamic biological networks and their fluctuations in response to various stimuli. It fundamentally governs human health and disease rather than just to snapshot a temporal time point of this sophisticated system. The combination of single cell technique and omics science takes full advantage of both techniques, and gets rid of their disadvantages. Single-cell based omics data reflect dynamic systemic cellular biology (from genes, and proteins to their downstream representations such as cellular metabolites), responding to various stimuli. Fortunately, technological improvement has made it feasible to realize the omics investigation based on single cell level. Here we focus on the whole process of single cell experiment in human and animals, isolation technique of single cell, experimental process, and potential clinical applications.

12.1 Technique for Single-Cell Isolation

The isolation of single cells from a solid tissue mainly needs two steps. First, the tissue is dissected from the tissue and dissociated into individual cells, usually by enzymatic disaggregation. Second, single cells are moved into individual reaction chambers for lysis and further processing. There are several techniques for single cell isolation, including serial dilution, micromanipulation (mechanical-using a simple mouth pipette and optimal), fluorescence-activated cell sorting (FACS), laser-capture microdissection (LCM) and microfluidic devices.

12.2 Serial Dilution

The simplest method to obtain single cell from heterogeneous populations is probably to serially dilute a sample solution until only one cell remain in ‘certain volume’ of culture medium, then equal divided the cells suspensions into small well plate in the same volume, which was firstly used for bacteria investigation (Schut et al. 1993; Button et al. 1993). Recently it has been successfully used for the single cell isolation of colon stem cell and acute leukemia cell by some groups (Reizel et al. 2011; Shlush et al. 2012). This method is simple and easy to perform, and maybe suitable for the isolation of the most abundant cells. However the isolation of rare cells is not feasible using this way, unless the target cells are enriched. Although the use of microtiter plates improved this method's efficiency (Connon and Giovannoni 2002), this single-cell isolation technique is stochastic and has the uncertainness of single cells in the individual volume wanted (e.g. maybe there are zero or more than one cells in each well). We used this method to isolate the primary lung cancer cell and found that it is hardly to control the quality and quantity.

12.3 Micromanipulation – Simple Mouth Pipette

As micromanipulation methods are easy and cheap, they are the most commonly used single-cell isolation methodologies. There are two types of micromanipulation: mechanical and optical micromanipulation (Brehm-Stecher and Johnson 2004), while they are only applicable to cells in suspension, low-throughput, and susceptible to errors, such as misidentification of a cell under a microscope. These disadvantages are partially addressed by semi-automated devices for cell isolation, with which an expert operator can isolate approximately 50–100 cells per hour. Here we discuss how to isolate single primary lung cancer cell with the simple mouth pipette in detail.

12.4 Micromanipulation – Protocol

- A set of simple mouth pipette, melting point capillary glass tube (0.9–1.1 × 120 mm), alcohol burner, a stereomicroscope, as shown in Fig. 12.1.
- Purification and proliferation of primary lung cancer cells and single-cell plating

Primary lung cancer cells were isolated from patient lung tissue with the modification as described by J. Seo et al. (2013). Briefly, the lung cancer sample tissues were sectioned to 40–100 um and then cultured on Matrigel-coated culture dishes. The tissue sections were treated with dispase to detach the desired population of cells, which were then cultured in N2-supplemented medium. The remnant fibroblasts were discarded because adhesion and proliferation were inhibited by using serum-free medium. After the initial growth of cancer cells, basal medium (DMEM) containing 10 % (v/v) FBS and cytokines, BEI (10 ng bFGF/ml, 10 ng EGF/ml, 10 ng IGF/ml), was used to enhance development. The isolated lung cancer cells were then plated by one single cell into each well of a 96 wells plate in basal medium (DMEM) containing 10 % (v/v) FBS and cytokines, BEI (10 ng FGF/ml, 10 ng EGF/ml, 10 ng IGF/ml) with a simple mouth pipette as previous description (Zong et al. 2012; Kurimoto et al. 2007).

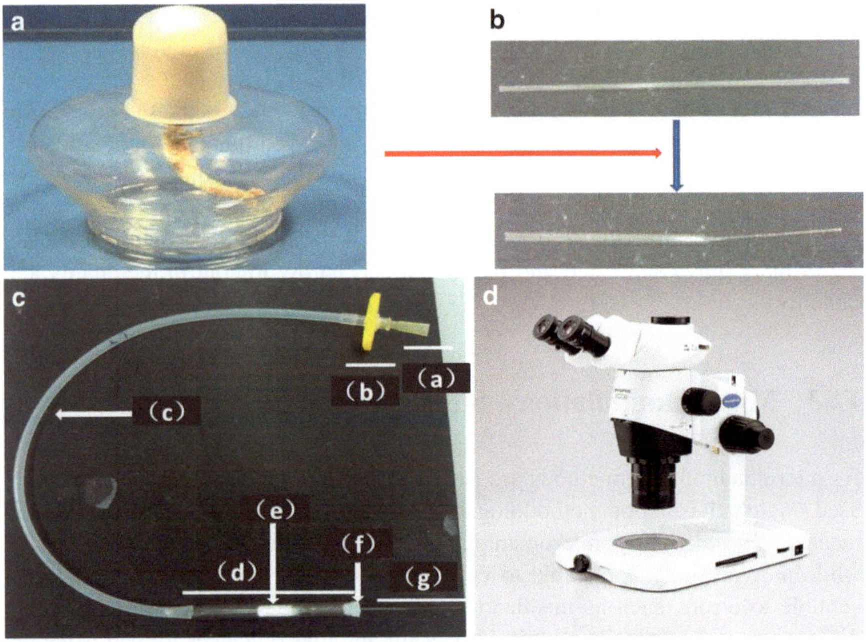

Fig. 12.1 Instruments for single cell isolation. (**a**) Alcohol burner; (**b**) A melting point capillary glass tube (0.9–1.1*120 mm) was drew into a thinner tube; (**c**) A mouth pipette, (*a*) a 200 ul micropipette, (*b*) PVDF filter (0.45 μm), (*c*) silicone tube, (*d*) glass adapter for capillary glass tube and silicone tube, (*e*) adsorbent cotton, (*f*) silicone adapter for capillary glass tube, (*g*) capillary glass tube; (**d**) Olympus SZX16 stereomicroscope

- Single cell culture and maintenance

About 57 % (55/96) wells were successfully plated with one single cell in each after 24 h of the cell plating. Then pictures of each well with the selected single cell were record every 3 days, as shown in Figs. 12.2, 12.3, 12.4, 12.5, 12.6, 12.7, 12.8 and 12.9. About 60 % of selected single cells (33/55) wells could successfully proliferate into one lung cancer cell clone. All the pictures were taken with 200× lens exception of marked with the magnification ×100.

12.5 Summary

The aim of our initial study was to develop an effective procedure and protocol to isolate, purify, and culture patient-derived lung cancer single cell for use in rapid anti-carcinogenic drugs creening applications. The carcinoma tumor samples were categorized as adenocarcinoma in the performed experiment. We have developed a procedure for culturing pure populations of primary NSCLCs in vitro by enzyme

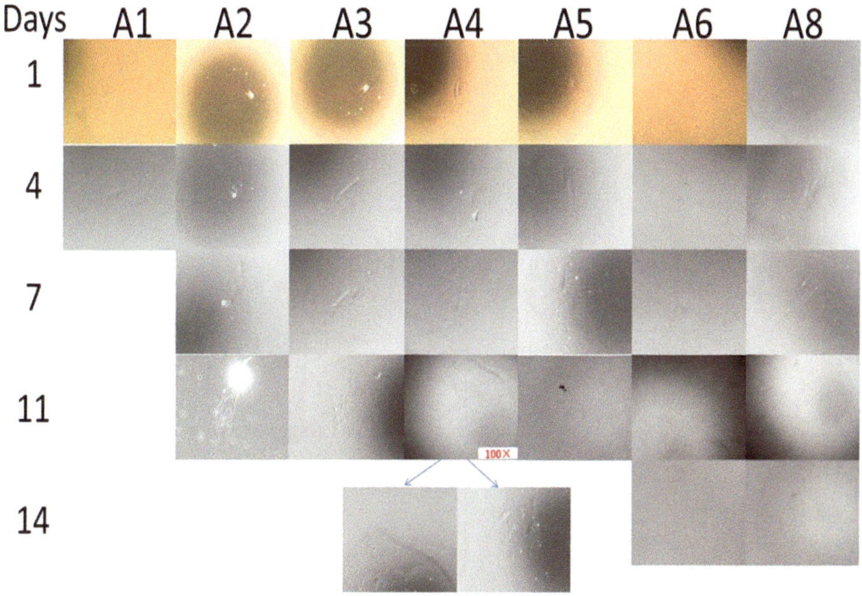

Fig. 12.2 Photos of single lung cancer cell in A row wells of the 96 well, where single cell was successfully plated

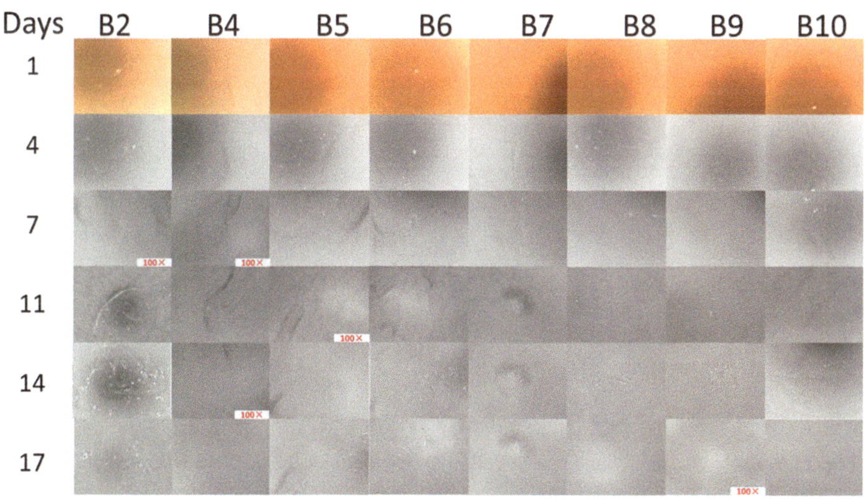

Fig. 12.3 Photos of single lung cancer cell in B row wells of the 96 well, where single cell was successfully plated

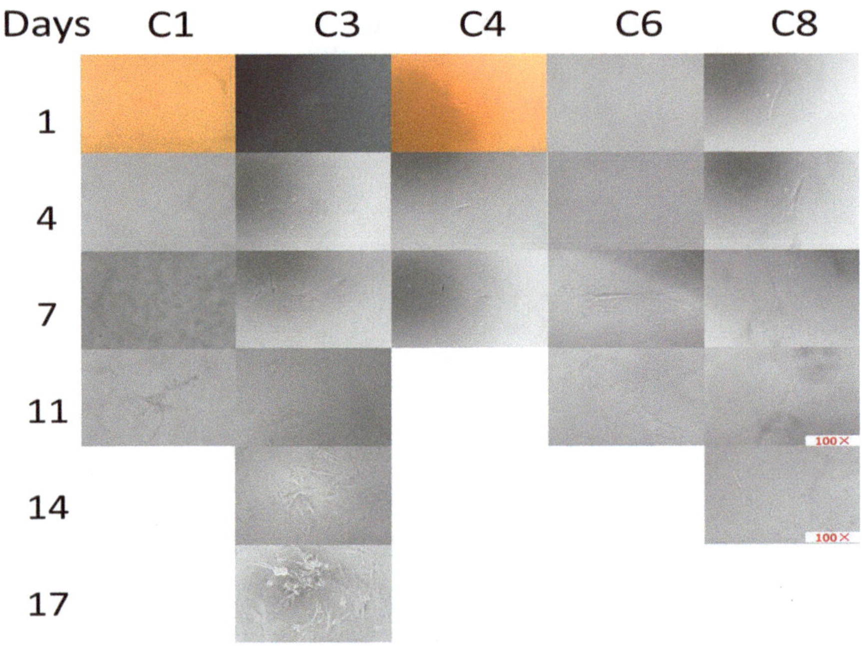

Fig. 12.4 Photos of single lung cancer cell in C row wells of the 96 well, where single cell was successfully plated

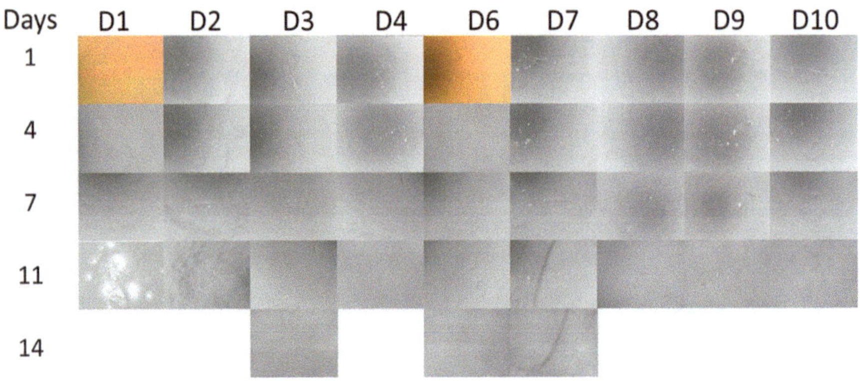

Fig. 12.5 Photos of single lung cancer cell in D row wells of the 96 well, where single cell was successfully plated

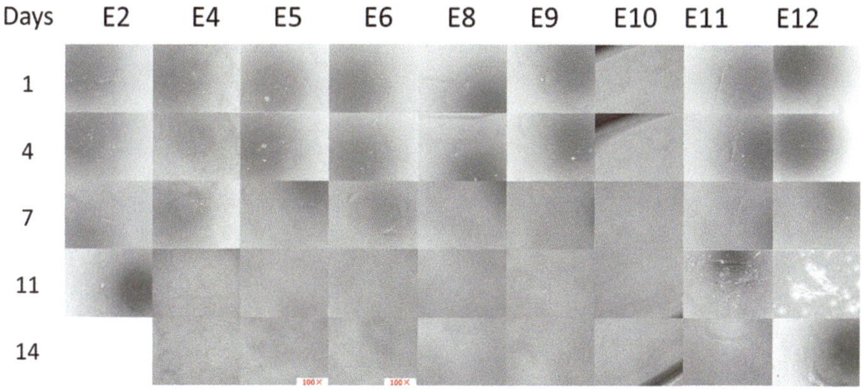

Fig. 12.6 Photos of single lung cancer cell in E row wells of the 96 well, where single cell was successfully plated

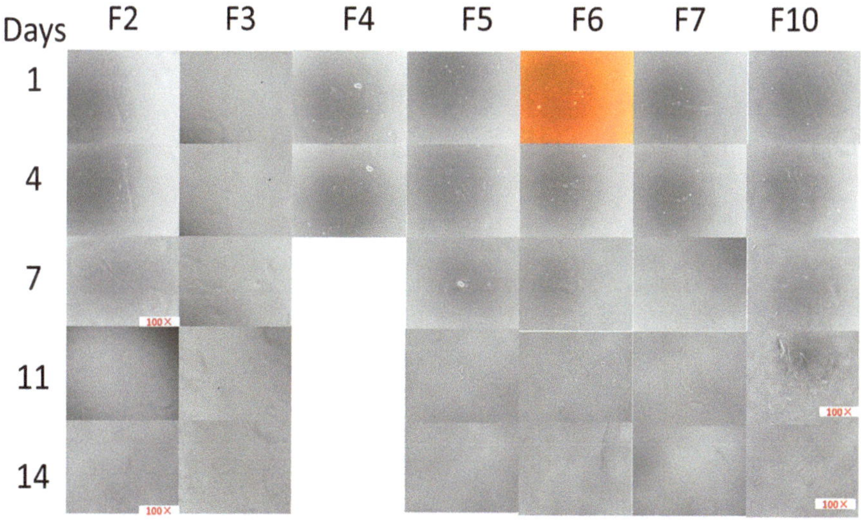

Fig. 12.7 Photos of single lung cancer cell in F row wells of the 96 well, where single cell was successfully plated

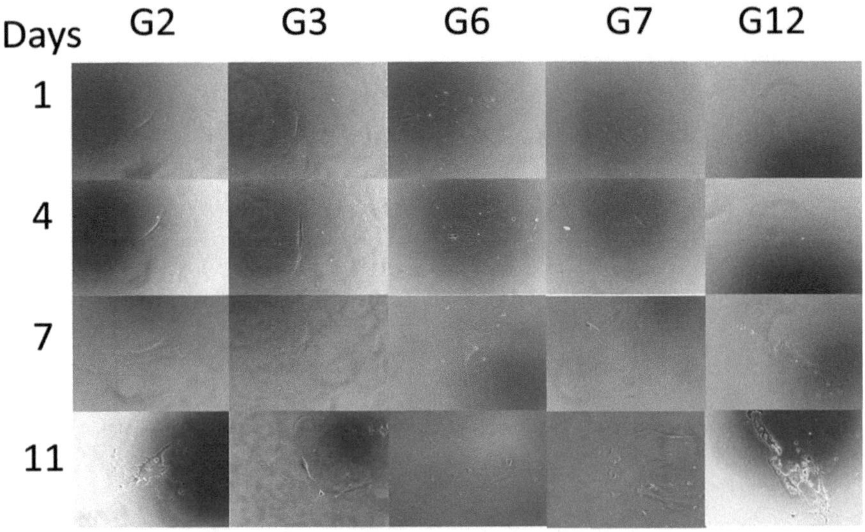

Fig. 12.8 Photos of single lung cancer cell in G row wells of the 96 well, where single cell was successfully plated

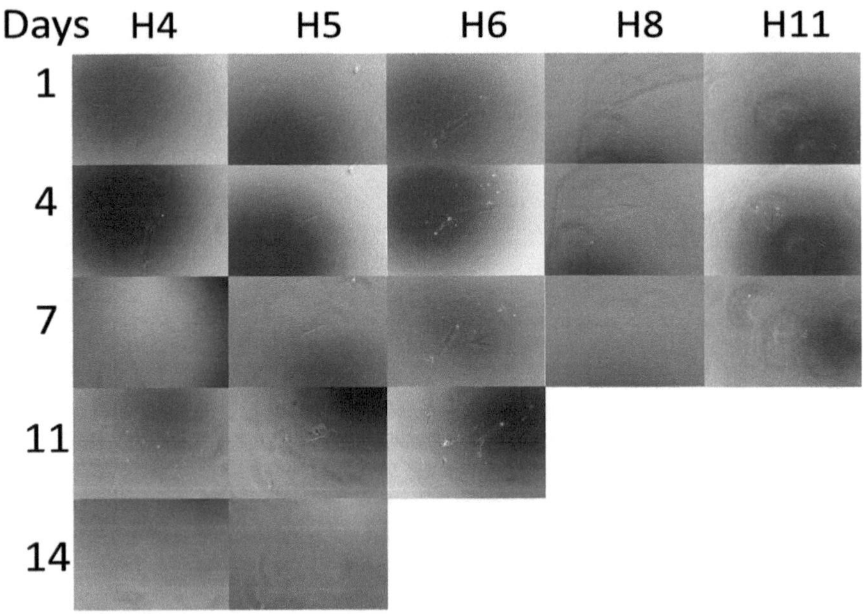

Fig. 12.9 Photos of single lung cancer cell in H row wells of the 96 well, where single cell was successfully plated

dissociation, serum-free medium-based purification, and growth factors-induced proliferation. The method is simple and efficient and converts 40–50 % of tumor samples to cultured cancer cells within 4 weeks. Such method was developed to facilitate anti-cancer drug development, rapid screening processes and further characterization studies.

References

Brehm-Stecher B, Johnson EA. Single-cell microbiology: tools, technologies, and applications [J]. Microbiol Mol Biol Rev. 2004;68(3):538–59.

Button D, Schut F, Quang P, et al. Viability and isolation of marine bacteria by dilution culture: theory, procedures, and initial results [J]. Appl Environ Microbiol. 1993;59(3):881–91.

Connon SA, Giovannoni SJ. High-throughput methods for culturing microorganisms in very-low-nutrient media yield diverse new marine isolates [J]. Appl Environ Microbiol. 2002;68(8):3878–85.

Frumkin D, Wasserstrom A, Kaplan S, et al. Genomic variability within an organism exposes its cell lineage tree [J]. PLoS Comput Biol. 2005;1(5):e50.

Kurimoto K, Yabuta Y, Ohinata Y, et al. Global single-cell cDNA amplification to provide a template for representative high-density oligonucleotide microarray analysis [J]. Nat Protoc. 2007;2(3):739–52.

Reizel Y, Chapal-Ilani N, Adar R, et al. Colon stem cell and crypt dynamics exposed by cell lineage reconstruction [J]. PLoS Genet. 2011;7(7):e1002192.

Schut F, de Vries EJ, Gottschal JC, et al. Isolation of typical marine bacteria by dilution culture: growth, maintenance, and characteristics of isolates under laboratory conditions [J]. Appl Environ Microbiol. 1993;59(7):2150–60.

Seo J, Park S-J, Kim J, et al. Effective method for the isolation and proliferation of primary lung cancer cells from patient lung tissues [J]. Biotechnol Lett. 2013;35(8):1165–74.

Shlush LI, Chapal-Ilani N, Adar R, et al. Cell lineage analysis of acute leukemia relapse uncovers the role of replication-rate heterogeneity and microsatellite instability [J]. Blood. 2012;120(3):603–12.

Zong C, Lu S, Chapman AR, et al. Genome-wide detection of single-nucleotide and copy-number variations of a single human cell [J]. Science. 2012;338(6114):1622–6.

Ding Zhang Dr. Zhang works as a fellow of Respiratory and critical care Medicine at Jiaotong University, Shanghai, China, and further his doctoral programme after Professor Xiangdong, Wang, at department of pulmonary diseases, Zhongshan Hospital, Fudan University, China. His main research is focused on the experimental and clinical research of mechanisms and stem cells therapy combination with telocytes of acute lung injury; he is also interested in lung cancer basic and clinical research and the interaction between lung cancer cells and the immune system; meanwhile, he did some primary single-cell-based lung cancer research as well.

Xiangdong Wang Dr. Wang works as a distinguished professor of Respiratory Medicine at Fudan University, China, adjunct professor of Molecular Bioscience at North Carolina State University, USA, and adjunct professor of Clinical Bioinformatics at Lund University. He serves as a Director of Biomedical Research Center, Fudan University Zhongshan Hospital. He was also pointed as visiting, guest and honor professors in Harbin Medical University (China), Zhejiang University (China), Kyoto Prefectural University of Medicine (Japan), and National Academy of Medicine (Romania). Dr. Wang acts as an Editor-in-Chief of Clinical & Translational Medicine, Journal of Clinical Bioinformatics and Journal of Epithelial Biology & Pharmacology, and Asian Editor of Journal of Cellular & Molecular Medicine. His main research is focused on the role of epithelial cells in the development of organ dysfunction, disease-specific biomarkers, and potential therapies.

Chapter 13
Cancer Bioinformatics

Mingyan He, Li Feng, and Jinglin Xia

Cancer bioinformatics is applying "informatics" techniques derived from disciplines such as computer science, applied mathematics, informatics, and statistics to understand and organize the information associated with cancer biology. The large-scale and multilevel omics-based biology researches, including genome, transcriptome, proteome, metabolome, phenome, and interactome, are the main contents of cancer bioinformatics in the age of post genomics.

With increasing evidence that the interaction and network between genes and proteins play a critical role in investigation of cancer molecular mechanisms, the single research on genome or proteome is hard to interpret the whole cancer biological behavior systematically and completely. Therefore, it is necessary and important to introduce a concept of "Systems Biology" into cancer research, to integrate genomics, proteomics, epigenetics, metabolomics, and interferomics to improve diagnosis, therapies and prognosis of cancer. Systems biology is an interdisciplinary field of study that focuses on complex interactions of all biological components within biological systems. Advances in cancer bioinformatics lead to large increases in the data favorable for study of systems biology. In 2005, The Cancer Genome Atlas (TCGA) was launched by National Institutes of Health (NIH), National Cancer Institute (NCI) and National Human Genome (NHGRI).

M. He
Liver Cancer Institute, Zhongshan Hospital, Fudan University, Shanghai, China
e-mail: hemingyan2011@163.com

L. Feng (✉)
Minhang Hospital, Fudan University, Shanghai, China
e-mail: fengweihong66666@sina.cn

J. Xia M.D., Ph.D. (✉)
Liver Cancer Institute, Zhongshan Hospital, Fudan University, Shanghai, China

Minhang Hospital, Fudan University, Shanghai, China
e-mail: xiajinglin@fudan.edu.cn

X. Wang (ed.), *Single Cell Sequencing and Systems Immunology*,
Translational Bioinformatics 5, DOI 10.1007/978-94-017-9753-5_13

TCGA's principal aims are to generate, control, merge, analyze, and interpret molecular profiles at the DNA, RNA, protein, and epigenetic levels for hundreds of clinical tumors from various tumor types and their subtypes. Some of the TCGA's works were published on the journal of Science in 2013. Molecular aberrations at different levels were discovered in large numbers of human tumors. The resulting rich data provide a major opportunity to develop an integrated landscape of complex molecular across tumor lineages and contribute to precision diagnosis, pharmaceutical development, and individualized therapy.

Cancer bioinformatics plays significant roles in the battle against human tumors. Firstly, cancer bioinformatics provides new strategies of biomarkers. Based on multiple genetic, protein or physic variables in cancer and high-throughout technologies, biomarkers were investigated from a single one to multiple markers, from the expression to function, and from the network to dynamic network. Selected tumor-specific networks and dynamic networks of genes or proteins in patients are correlated with early-diagnosis, development, response to therapy and prognosis of tumor, to identify and validate tumor-specific biomarkers. Secondly, pharmacogenomics is an important part of bioinformatics. It aims to develop rational means to optimize drug therapy and ensure maximum efficacy with minimal adverse effects by dealing with the influence of acquired and inherited genetic variation on drug response in patients, and develop new drugs using genomic information. In addition, cancer bioinformatics increases the efficiency and effectiveness of the precision medicine, which provides the safest and most effective therapeutic strategies based on genetic variations of each subject. In view of molecular network characters of each patient's tumor, cancer bioinformatics is expected to delivering the right treatment to the right patient in the right time through therapy design.

Mingyan He, Ph.D. Candidate of Hepatology, Fudan University. She got the Bachelor Degree in Clinical Medicine, Medical School of Nanchang University. Dr He did her internship in The First Affiliated Hospital of Nanchang University and won several outstanding student scholarships, honorable titles of excellent class, and honorable mention of Psychological drama. Dr He got Master Degree in Pediatrics from Shanghai Medical School of Fudan University, 2014.

Li Feng is a Professor and Director of Department of Science and Research, and Chief-in-Physician of Department of Internal Medicine, Minhang Hospital of Fudan University, Shanghai, China. She also serves as the General Secretary and Committee Member of Shanghai Medical Association of Digestive Endoscopy Committee, Digestive Endoscopy Esophageal Division and, Shanghai Traditional Chinese and Western Medical Association Digestion Professional Committee, Dr Feng won the "Shanghai Medical Science and Technology Award", "Minhang District Science and Technology Achievement Award" and others. She published more than 50 articles, acted as an Editor-in-Chief of Medical Monographs 2, and authored 8 books. Dr Feng also won the honorable title of "Shanghai Leading Talents", "Shanghai Workers of Outstanding Innovative Technology Achievement Awards", "Shanghai Medical Women Association Outstanding Female Physician Award.

Jinglin Xia Project Director. Dr. Xia is the vice president of Medical College, Fudan University, Shanghai, China. He finished postdoctoral research at the Medical Center of Pittsburgh University in 2002–2003. He serves as the editor of a number of international medical journals, executive member of the International Association of translational medicine; vice-chairman of the Shanghai Medical Association. Dr. Xia's main research focuses on tumor angiogenesis and tumor inflammatory microenvironment. Dr. Xia has published over 60 papers and plenty of books covering tumor angiogenesis, microRNAs, and inflammatory factors and so on.

MIX
Papier aus verantwortungsvollen Quellen
Paper from responsible sources
FSC® C105338

If you have any concerns about our products,
you can contact us on
ProductSafety@springernature.com

In case Publisher is established outside the EU,
the EU authorized representative is:
Springer Nature Customer Service Center GmbH
Europaplatz 3, 69115 Heidelberg, Germany

Printed by Libri Plureos GmbH
in Hamburg, Germany